AN LEABH

CC h

Current Topics in Eye Research

Contributors to This Volume

D. J. Chiarandini
J. Davidowitz
Keith Green
Elizabeth D. Hay
H. L. Kern
Thomas F. Linsenmayer
Giovanni Maraini
James L. Rae
Robert L. Trelstad
Klaus von der Mark

Current Topics in Eye Research

VOLUME 1

Edited by

JOSE A. ZADUNAISKY

Departments of Physiology and Ophthalmology
New York University Medical Center
School of Medicine
New York, New York

HUGH DAVSON

Department of Physiology
King's College
University of London
London, England

ACADEMIC PRESS
New York San Francisco London 1979
A Subsidiary of Harcourt Brace Jovanovich, Publishers

ACADEMIC PRESS, INC.
111 Fifth Avenue, New York, New York 10003

United Kingdom Edition published by
ACADEMIC PRESS, INC. (LONDON) LTD.
24/28 Oval Road, London NW1 7DX

LIBRARY OF CONGRESS CATALOG CARD NUMBER: 78–4816

ISBN 0–12–153001–9

PRINTED IN THE UNITED STATES OF AMERICA
79 80 81 82 83 84 9 8 7 6 5 4 3 2 1

Contents

LIST OF CONTRIBUTORS ix
PREFACE xi

Origin and Distribution of Collagens in the Developing Avian Cornea

ELIZABETH D. HAY, THOMAS F. LINSENMAYER, ROBERT L. TRELSTAD, AND KLAUS VON DER MARK

I. Introduction 1
II. Collagen Types 2
III. Procollagens 5
IV. Spatial Localization of Collagen Synthesis and Secretion 6
V. Aggregation of Collagen into Fibrils 10
VI. Cells Involved in the Synthesis of Corneal Collagen . 12
VII. Biochemical Studies of Collagen Types in the Developing Avian Cornea 21
VIII. Immunohistochemical Evaluation of Corneal Collagen Heterogeneity 25
IX. Summary 27
References 31

The Electrophysiology of the Crystalline Lens

James L. Rae

I. Introduction 37
II. Electrophysiologic Morphology 38
III. Extracellular Space 49
IV. Electrophysiology 54
V. Summary 84
References 85

Structure and Function of Extraocular Muscle Fibers

D. J. Chiarandini and J. Davidowitz

I. Structural Characteristics 91
II. Physiological Properties 107
References 140

Binding Proteins for Retinol in Retina and Pigment Epithelium

Giovanni Maraini

I. Introduction 143
II. The Transport Mechanism for Vitamin A in Plasma 145
III. The Role of the Pigment Epithelium in the Metabolism of Retinol within the Eye 149
IV. The Interaction of the Retinal Pigment Epithelium with Circulating Retinol-Binding Protein and the Uptake of Retinol 153
V. Cellular Retinol-Binding Proteins in Extraocular Tissues 158
VI. Binding Proteins for Retinol in Retinal Pigment Epithelium 160
VII. Binding Proteins for Retinol in the Retina and Photoreceptor Cell Fractions 167
VIII. Considerations and Summary 169
References 172

The Ocular Effects of Cannabinoids

KEITH GREEN

I. Cannabinoid History 175
II. Cannabinoid Chemistry 178
III. Clinical Ophthalmic Tests 181
IV. Preclinical Tests 185
V. Cannabinoid Pharmacology 193
VI. Local and Central Action 202
VII. New Cannabinoids 204
VIII. Future Investigations 210
References 212

Transport of Organic Solutes in the Lens

H. L. KERN

I. Introduction 217
II. Amino Acids 219
III. Sugars and Polyols 222
IV. Substituents of Nucleic Acids 225
V. Effects of Differentiation 226
VI. Effects of Aging 227
VII. Effects of Cations, Tonicity, and Chemical Agents . 229
VIII. Effects of Cataractogenic Agents and Cataract . . . 234
IX. Concluding Remarks 237
References 238

SUBJECT INDEX 241

List of Contributors

Numbers in parentheses indicate the pages on which the authors' contributions begin.

D. J. CHIARANDINI (91), Department of Ophthalmology, New York University Medical Center, New York, New York 10016

J. DAVIDOWITZ (91), Department of Ophthalmology, New York University Medical Center, New York, New York 10016

KEITH GREEN (175), Departments of Ophthalmology and Physiology, Medical College of Georgia, Augusta, Georgia 30902

ELIZABETH D. HAY (1), Department of Anatomy, Harvard Medical School, Boston, Massachusetts 02115

H. L. KERN (217), Department of Ophthalmology, Albert Einstein College of Medicine, Yeshiva University, Bronx, New York 10461

THOMAS F. LINSENMAYER (1), Developmental Biology Laboratory, Massachusetts General Hospital, Boston, Massachusetts 02114

GIOVANNI MARAINI (143), Department of Ophthalmology, University of Parma, 43100 Parma, Italy

JAMES L. RAE (37), Departments of Ophthalmology and Physiology and Biophysics, University of Texas Medical Branch, Galveston, Texas 77550

ROBERT L. TRELSTAD (1), Department of Pathology, Shriners Burns Institute and Massachusetts General Hospital, Boston, Massachusetts 02114

KLAUS VON DER MARK (1), Max-Planck-Institut für Biochemie, Abteilung Kühn, Martinsried, Germany

Preface

The publication of *Current Topics in Eye Research* was prompted by our recognition that authoritative summaries of progress in specific areas of basic and applied research in the rapidly expanding visual sciences are very much needed. Application of the basic sciences, including cell biology, physiology, biophysics, pharmacology, biochemistry, immunology, and pathology to eye research is undergoing a definite expansion as reflected by the increased volume of published research and by the greater number of scientists entering the field.

Because of the increased number of scientific contributions in these different disciplines, investigators have been called upon to become more specialized, and therefore a need for education in those areas outside of such specialization has been created.

This new serial publication meets this educational need by means of original articles that include up-to-date revisions and timely evaluations of the states of particular fields, including the scientific contributions of the authorities that have prepared each chapter.

Although the area of clinical ophthalmology usually is well covered by published review articles, the basic aspects of eye and vision research do not have a forum where the researcher, the academic investigator, or the interested ophthalmologist can find such information. This publication attempts to constitute such a forum in order to reflect the present status of specific areas of eye and vision research.

J. A. ZADUNAISKY
HUGH DAVSON

Origin and Distribution of Collagens in the Developing Avian Cornea

ELIZABETH D. HAY[1]
THOMAS F. LINSENMAYER[1,2]
ROBERT L. TRELSTAD[3]
KLAUS VON DER MARK[4]

I. Introduction . . . 1
II. Collagen Types . . . 2
III. Procollagens . . . 5
IV. Spatial Localization of Collagen Synthesis and Secretion . . . 6
V. Aggregation of Collagen into Fibrils . . . 10
VI. Cells Involved in the Synthesis of Corneal Collagen . . . 12
VII. Biochemical Studies of Collagen Types in the Developing Avian Cornea . . . 21
VIII. Immunohistochemical Evaluation of Corneal Collagen Heterogeneity . . . 25
IX. Summary . . . 27
References . . . 31

I. INTRODUCTION

The optical function of the cornea requires the coherent transmission of light through an extracellular matrix (ECM) which predominantly consists of highly ordered layers of collagen fibrils combined with glycosaminoglycans (GAG). Transparency is partly the result of the precise spatial arrangement of matrix macromolecules which is achieved during develop-

[1] Department of Anatomy, Harvard Medical School, Boston, Massachusetts.
[2] Developmental Biology Laboratory, Massachusetts General Hospital, Boston, Massachusetts.
[3] Department of Pathology, Shriners Burns Institute and Massachusetts General Hospital, Boston, Massachusetts.
[4] Max-Planck-Institut für Biochemie, Abteilung Kühn, Martinsried, Germany.

ISBN 0-12-153001-9

ment. Thus, corneal morphogenesis is of considerable interest to both ophthalmologists and developmental biologists. In the past decade, studies done chiefly in the embryonic chick have led to a characterization of the roles of the various corneal cell types in collagen and GAG production. In addition, the chemistry and spatial distribution of the constituent collagens have recently been clarified and some conclusions have been drawn concerning the morphogenetic forces that operate during the precise spatial deposition of these molecules. In the following review, a number of these recent studies of collagen biochemistry, ultrastructure, and immunohistochemistry will be presented, with emphasis on the formation and distribution of collagens in the developing avian cornea.

II. COLLAGEN TYPES

Collagen, a ubiquitous component of many different connective tissues, is a family of genetically distinct molecules. There are at least four and probably more such distinct molecules in vertebrates. The known collagen types I–III consist of three individual polypeptide chains, called α chains, each with a molecular weight of approximately 95,000. In all three collagen types, the component α chains have certain common properties including glycine as every third amino acid residue, high concentrations of proline, hydroxyproline, and hydroxylysine, and the ability to assemble with two adjacent α chains to form a triple helical collagen molecule (sometimes called the tropocollagen molecule; Fig. 1). The collagen molecules have the ability to self-assemble into various forms including native fibrils with the characteristic 67 nm cross striations which are found in tissues and segment-long-spacing (SLS) crystallites that can be formed *in vitro* (for review, see Gross, 1973).

Although most collagens have these properties that make them similar, the component α chains in the various types have different amino acid sequences (Fig. 2). Hence, the molecules represent different gene products. The most ubiquitous type of collagen molecule, designated type I collagen or $[\alpha 1(\mathrm{I})]_2\,\alpha 2$, has a chain distribution consisting of two $\alpha 1$ type I chains and one $\alpha 2$ chain. The $\alpha 1$ and $\alpha 2$ designations refer to the elution positions of the chains from carboxymethyl (CM)-cellulose columns where these chains were first noted and named (Piez *et al.*, 1963). Type I collagen was initially described in skin and bone (Piez *et al.*, 1963; Miller *et al.*, 1967, 1971) and now has been found as a major or minor component of most embryonic and adult connective tissues, including the avian and mammalian corneas (Trelstad and Kang, 1974; Katzman *et al.*, 1974).

In embryonic and adult cartilage the predominant collagen molecule

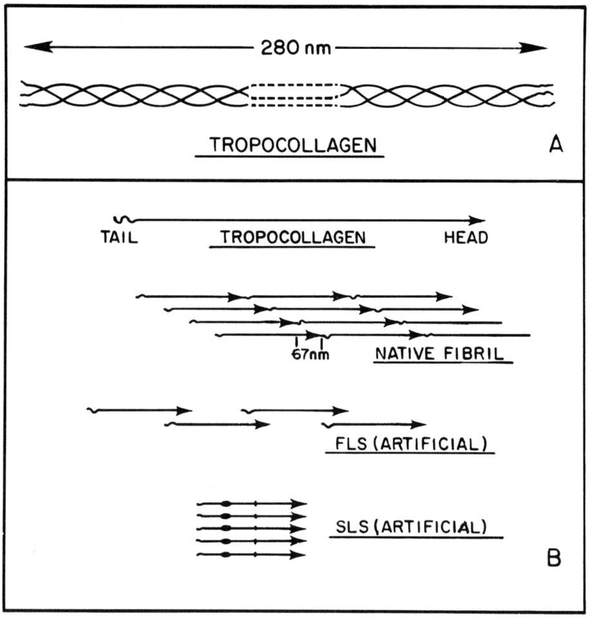

FIG. 1. (A) Diagram showing that the substructure of the tropocollagen molecule consists of three polypeptide chains (α chains) arranged in a triple helix. (B) The tropocollagen molecule in solution can be polymerized into quarter-staggered native fibrils in 0.1 *N* NaCl *in vitro*. Under different conditions, other patterns form, such as SLS (segment-long-spaced) when ATP is added, and FLS (fibrous-long-spaced) when serum glycoprotein is added. (After Gross, 1956.)

consists of three identical α chains, different in amino acid composition from that of skin and bone. This molecule is designated $[\alpha1(II)]_3$ or type II collagen (Miller and Matukas, 1969; Trelstad *et al.*, 1970). Although the $\alpha1$ type I and $\alpha1$ type II chains are different gene products, both elute in essentially the same position from CM-cellulose columns (see Figs. 13, 14), but they can be separated on an anion exchange resin (Trelstad *et al.*, 1972; Linsenmayer *et al.*, 1973a,b,c; Linsenmayer, 1974). Recently, type II collagen, initially thought to be diagnostic for cartilage, has been found in notochord (Linsenmayer *et al.*, 1973b; Miller and Matthews, 1974; von der Mark *et al.*, 1976a), vitreous body (Swann *et al.*, 1972; Newsome *et al.*, 1976; Smith *et al.*, 1976), and the primary embryonic chick corneal stroma (Linsenmayer *et al.*, 1977; von der Mark *et al.*, 1977).

At least two other collagen types have been characterized; the type III collagen (designated $[\alpha1(III)]_3$) of certain embryonic tissues (Epstein, 1974), blood vessels and leiomyomas (Chung and Miller, 1974; Trelstad, 1974), and the type IV collagen of basement membranes designated

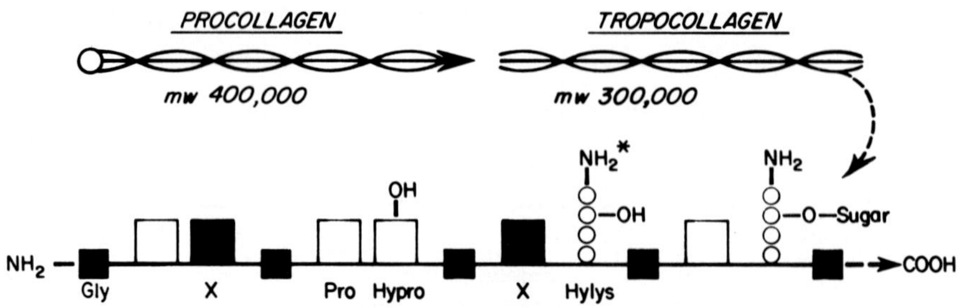

	INTERSTITIAL COLLAGENS				BASEMENT MEMBRANE		
	α1(I)	α2	α1(II)	α1(III)	TYPE IV	A	B
3OHPro	0	1	2	1	11	7	9
4OHPro	99	90	100	115	125	110	109
Pro	117	115	115	95	62	99	119
Gly	333	333	333	350	340	346	334
Ala	125	120	105	90	33	53	45
Hylys	5	5	20	7	45	23	37
Sugar (galactose, glucosylgalactose)	0.5%	1%	5%	1%	10%	2%	5%

*covalent crosslink

FIG. 2. (Top) In procollagen, nonhelical polypeptides are present on both ends of the α chains, which are cleaved off when the molecule is secreted. (Middle) The amino acid sequence of one of the α chains is shown in part. (Bottom) Partial amino acid composition of α chains of some collagens. Glycine (Gly) occurs in every third position (333/1000 residues), and proline (Pro) and hydroxyproline (OHPro) frequently in the other two positions (200–250/1000 residues). Of the remaining 450 amino acids, hydroxylysine (Hylys) is the most important, although outnumbered by alanine (Ala), glutamine, and others (X, middle diagram) (Grant and Prockop, 1972). In addition type III collagen and basement membrane (type IV) contain cystine (Chung and Miller, 1974; Kefalides, 1975). The α chains A and B are found in certain basement membranes (Chung *et al.*, 1976; Burgeson *et al.*, 1976); it is possible that they will be called by different names in the future.

$[\alpha 1(IV)]_3$ (for review see Kefalides, 1973, 1975). There is now reason to suspect the presence of yet other collagen types (Burgeson *et al.*, 1976; Chung *et al.*, 1976; Trelstad and Lawley, 1977), all of which may belong to the basement-membrane class of collagens (Fig. 2).

During the biosynthesis of the different types of collagens, a number of posttranslational modifications occur, including the hydroxylation of proline and lysine, the glycosylation of hydroxylysine, and oxidation of lysine and hydroxylysine to reactive aldehydes which then participate in the formation of crosslinks that function in stabilizing collagen fibrils, once formed. The absolute and relative amounts of these modifications vary among collagens from different sources (Fig. 2). Type II collagen from cartilage, for example, contains about three times as much hydroxylysine, and about ten times the hexose of skin collagen (Trelstad *et al.*, 1970). Bone collagen isolated from rachitic animals has about twice the hydroxylysine content of normal bone and skin collagen (Toole *et al.*, 1972). Although the function of the glycosylated residues in collagen remains largely unknown, they have been implicated as recognition sites during the platelet–collagen interaction that occurs as the primary event in hemostasis (Jamieson *et al.*, 1971; Kang *et al.*, 1974), and they have also been implicated as one of the factors influencing collagen fibrillogenesis. The evidence for the involvement of glycosylation in fibrillogenesis is derived largely from the work of Schofield *et al.* (1971), who showed that an inverse relationship exists between the diameter of fibrils and the amount of carbohydrate covalently bound to the collagen extracted from those tissues. Collagens with low carbohydrate content are found in tissues that contain fibrils with a wide range of diameters and a large mean diameter. This relationship also appears to hold true for corneal collagen, at least as examined in the adult (Schofield *et al.*, 1971; Cintron, 1974).

For further information on collagen types and collagen chemistry, the reader is directed to reviews by Gross (1973), Trelstad (1973), Miller and Matukas (1974), Fietzek and Kühn (1976), and Miller (1976).

III. PROCOLLAGENS

All four known types of collagens are synthesized as higher-molecular-weight precursor molecules termed procollagens (for reviews, see Bornstein, 1974; Martin *et al.*, 1975; Prockop *et al.*, 1976). The procollagens of type I from fibroblasts (Dehm and Prockop, 1971, 1972, 1973; Layman *et al.*, 1971; Smith *et al.*, 1972; Goldberg *et al.*, 1972) and calvaria (Monson *et al.*, 1975; Fessler *et al.*, 1973, 1975) and procollagens of type II from cartilage (Dehm and Prockop, 1973) are subsequently converted to the lower-molecular-weight tropocollagen molecules which are then deposited in the extracellular matrix. The type IV collagen, from certain basement membranes, at least, remains in the tissue in the form of unprocessed procollagen (Minor *et al.*, 1976).

A detailed model for the structure of the procollagen and the sequential steps required for its conversion into final form is gradually being established. The procollagen molecule is composed of three pro-α chains and contains a central triple helical portion, with shorter nonhelical extension pieces at both the NH_2 and COOH terminal ends that contain cysteine residues (Furthmayr *et al.*, 1972; von der Mark and Bornstein, 1973; Tanzer *et al.*, 1974; Uitto *et al.*, 1976). In type I procollagen, however, only the COOH terminal extension contains *inter*chain disulfide crosslinks (Fessler *et al.*, 1975; Byers *et al.*, 1975), whereas in type III procollagen both extensions are linked by disulfide bonds (Church *et al.*, 1973; Byers *et al.*, 1974; Lichtenstein *et al.*, 1975; Nowack *et al.*, 1976b). Outside the cell, procollagen is converted to tropocollagen by limited cleavage of portions of the nonhelical NH_2 and COOH terminal extensions by procollagen peptidases (Lapiere *et al.*, 1971; Bornstein *et al.*, 1972). The converted molecules are then polymerized in the extracellular matrices to form collagen fibrils.

It should be noted that the completely converted collagen molecules, as found in extracellular fibrils, still have at both ends short nonhelical portions termed telopeptides, which carry both crosslinking sites and major antigenic determinants. Those terminal determinants, which are pepsin-labile, have been precisely located in type I collagen on the cyanogen bromide peptides α1(I)-CB(0,1) and in α2-CB1 at the amino end, and on α1(I)-CB6 and on α2-CB5 and the carboxy end. With the use of the passive hemagglutination technique, helical, conformation-dependent antigenic sites and central, sequence-dependent determinants have also been characterized (for review see Timpl, 1976; Furthmayr and Timpl, 1976; and Fig. 3). Terminal antigenic sites are mostly recognized when rabbits are immunized with collagen, whereas helical determinants are recognized by most other animals. Collagen is a rather weak immunogen, but antisera of high titers that do not cross-react with the triple helical part of the procollagen molecule can be raised against the N-terminal (Timpl *et al.*, 1973; von der Mark *et al.*, 1973) and C-terminal extension peptides of procollagen (Sherr and Goldberg, 1973; Dehm *et al.*, 1974).

IV. SPATIAL LOCALIZATION OF COLLAGEN SYNTHESIS AND SECRETION

Immunohistochemistry and autoradiography have made it possible to localize some of the intracellular steps of collagen synthesis within cell organelles. Using procollagen-specific antibodies linked to ferritin, it has been possible to localize procollagen by electron microscopy in the lumen of the endoplasmic reticulum (Fig. 4) and within Golgi vacuoles (Fig. 5) of tendon and corneal fibroblasts (Olsen and Prockop, 1974; Olsen *et al.*, 1975; Nist *et al.*, 1975). Olsen *et al.* (1973b) also localized proline hy-

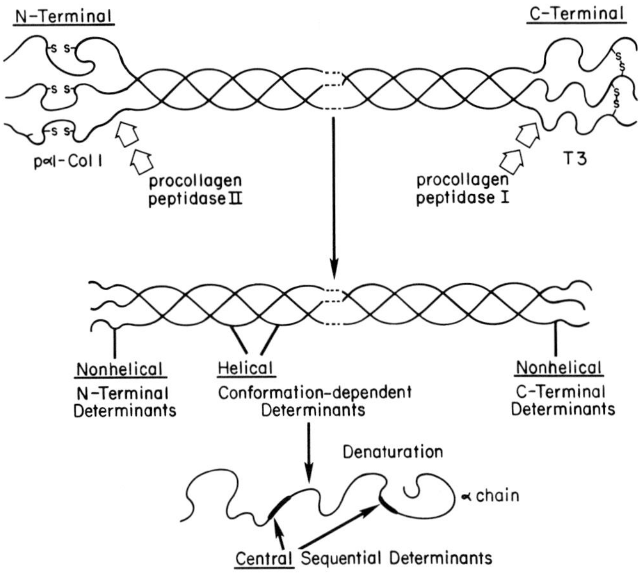

FIG. 3. Procollagen carries antigenic sites at both the N- and the C-terminal extensions (top). They are sensitive to reduction of disulfide bonds under denaturing conditions (Dehm *et al.*, 1974; Timpl, 1976). Conversion of procollagen to collagen by procollagen peptidases leaves short, nonhelical sequences on both the ends, which carry major antigenic sites. Terminal determinants can be removed by controlled proteolytic degradation. Native collagens also contain conformation-dependent antigenic determinants in the helical part of the molecule which are lost upon denaturation (helical determinants). Antigenic determinants in the central part that are active after denaturation are sequence-dependent (central determinants). (After Timpl, 1976.)

droxylase with ferritin-labeled antibodies within the lumen of the endoplasmic reticulum.

Autoradiographic studies of collagen synthesis using ^{3}H-proline as a precursor indicate that the granular endoplasmic reticulum alone is involved in the manufacture of collagen; much of the ^{3}H-proline-labeled protein subsequently translocates to Golgi vesicles (Revel and Hay, 1963; Hay and Revel, 1963; Ross and Benditt, 1965; Salpeter, 1968; Weinstock and Leblond, 1974). Thus, collagen is probably synthesized by messenger RNA on the ribosomes of the endoplasmic reticulum, accumulates within the lumen of the reticulum where it is hydroxylated, and then moves into the Golgi apparatus. The Golgi vacuoles are transported, in some manner dependent on microtubules (Ehrlich *et al.*, 1974), to the cell surface where the product is discharged.

The function of the Golgi apparatus seemingly has to do with the packaging of collagen for secretion from the cell. The overall conformation of the

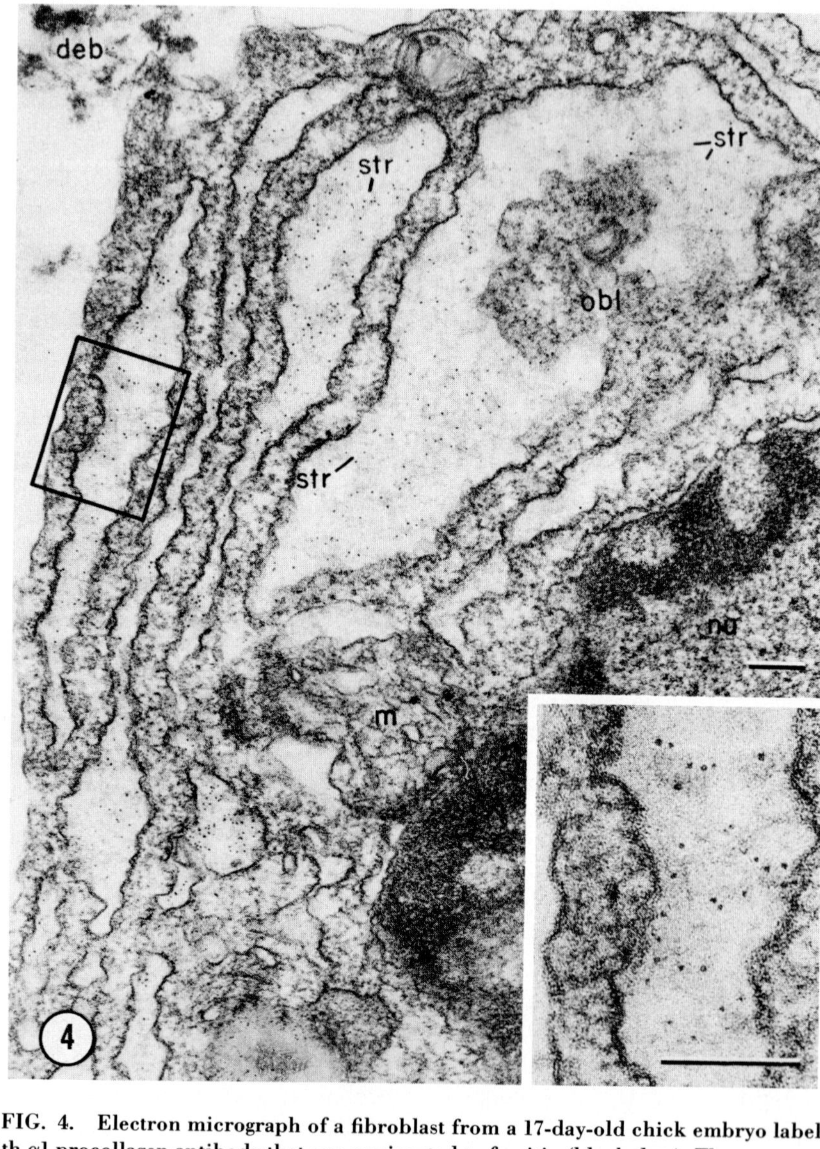

FIG. 4. Electron micrograph of a fibroblast from a 17-day-old chick embryo labeled with $\alpha 1$ procollagen antibody that was conjugated to ferritin (black dots). The contents of the endoplasmic reticulum are aggregated into strands (str) that contain procollagen. At obl, the reticulum is sectioned obliquely. The inset shows the area in the rectangle at higher magnification. (m) Mitochondrion; (deb) debris; (nu) nucleus. × 57,750. Inset, × 146,300. Bar represents 0.1 μm. (From Nist *et al.*, 1975, by courtesy of the Rockefeller University Press.)

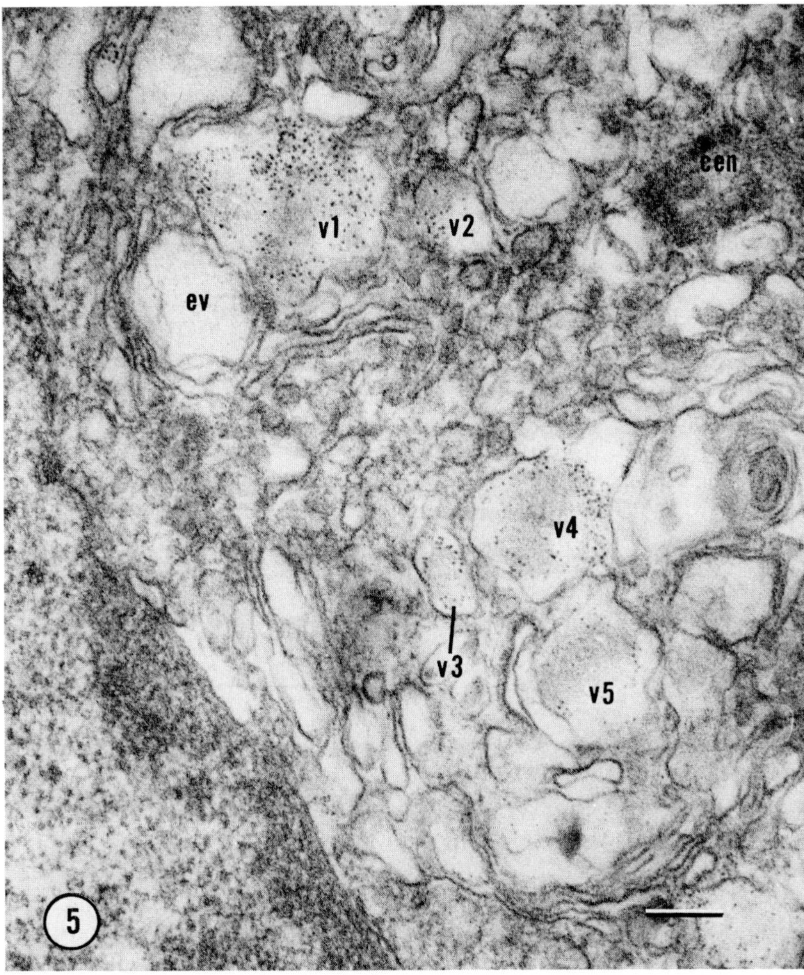

FIG. 5. Electron micrograph of the Golgi region of fibroblast labeled with $\alpha 1$ procollagen antibody. Both corneal and tendon fibroblasts show the same pattern of labeling. The Golgi vacuoles (labeled v1 to v5) contain a finely filamentous material which is rich in procollagen. The empty vacuole (ev) probably lost its content during tissue preparation; (cen) centriole. $\times$ 84,700. Bar represents 0.1 μm. (From Nist *et al.*, 1975, by courtesy of the Rockefeller University Press.)

collagen molecules polymerized within Golgi vacuoles may be cigarlike (Trelstad, 1971), rodlike (Revel and Hay, 1963), threadlike (Weinstock and Leblond, 1974), or filamentous (Hay and Dodson, 1973; Nist *et al.*, 1975). The Golgi apparatus manufactures glycosaminoglycans (GAG), including chondroitin sulfate (Revel, 1970; Young, 1973), which in appropriately stained sections can be seen to combine in the Golgi vacuoles with the fibrous molecules presumed to be collagen (Revel and Hay, 1963; Revel, 1964). Thus, it seems likely that collagen is secreted from the cell in partially assembled collagen–GAG packets (Trelstad *et al.*, 1974).

V. AGGREGATION OF COLLAGEN INTO FIBRILS

Purified preparations of collagen types I, II, and III when placed in physiological conditions of pH, ionic strength, and temperature will spontaneously self-assemble into native 67 nm cross-striated fibrils. The collagen monomer itself thus has information residing in the spatial distribution of charged and hydrophobic regions along its axis sufficient to allow it to spontaneously associate with other molecules into a highly ordered fibrillar aggregate which is indistinguishable from those formed *in vivo* (Wood, 1964; Gross and Kirk, 1958).

In vitro studies of steps in collagen fibril formation (Trelstad *et al.*, 1976) indicate that the monomer–polymer transition may be a multistage process involving distinct intermediate aggregates or subassemblies. The initial aggregate is 4 nm wide, about the width of the hypothesized collagen microfibril (Miller and Parry, 1973), and greater than 30 nm long, the length of the collagen molecule. These initial aggregates appear to assemble into larger tactoidal structures by lateral association. The tactoidal intermediates form linear aggregates that seem to entwine with each other in a ropelike fashion to form the final fibril (Trelstad *et al.*, 1976). *In vivo*, the forming fibril may be closely associated with the outer face of the cell in some tissues (Trelstad *et al.*, 1974, 1976) and the amount and type of associated GAG may influence final fibril size (Toole and Lowther, 1968a,b).

The precise spatial distribution of the collagen in the corneal stroma is a prime determinant of its transparency. In the corneal stroma of both mammalian and submammalian species the collagen fibrils are uniform in diameter (~ 25 nm) in the embryo and the adult, and are packed with a uniform distance between the fibrils (Hay and Revel, 1969; Hart and Farrell, 1969; Trelstad and Coulombre, 1971). In submammalian corneas in addition to the precise interfibrillar spacing of collagen within a stromal layer, adjacent layers are oriented at approximately 90° to each other.

Moreover, as the stroma is visualized from the epithelial to the endothelial surface, this orthogonal set of layers undergoes a progressive clockwise shift such that the entire stroma describes a spriraling configuration of over 200° total displacement (Coulombre, 1965; Trelstad and Coulombre, 1971; Fig. 6).

This unusual spiraling orthogonal pattern is bilaterally asymmetrical in that the shift is clockwise in both eyes. The pattern resembles a cholesteric

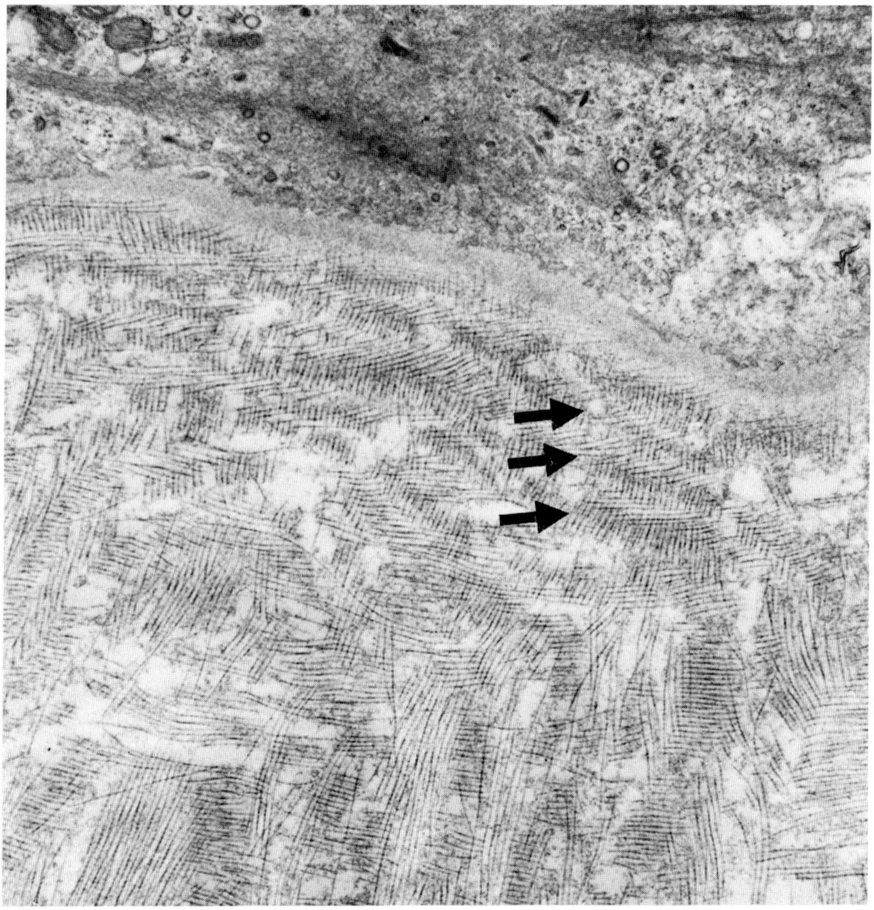

FIG. 6. Electron micrograph of stage 34 (8-day-old) embryonic chick cornea sectioned parallel to the corneal surface. The collagenous layers of the subepithelial primary stroma (arrows) are apparent because of corneal curvature. The collagen fibrils within a layer are spaced uniformly distant from each other and orthogonal to those in adjacent layers. The interfibrillar distance (60 nm) within a layer is the same as the macroperiod of the striated collagen fibril. × 13,500. (From Trelstad and Coulombre, 1971, by courtesy of the Rockefeller University Press.)

liquid crystal, and presumably forces involved in macromolecular self-assembly are operative in the generation of this configuration. Interestingly, in studies of the *in vitro* polymerization of collagen, patterns of othogonal and spiral nature spontaneously form *in vitro* (Trelstad *et al.*, 1976). Although considerable order can be generated by purified collagen solutions *in vitro*, it nonetheless seems likely that cellular influences operate particularly at the site of fibril assembly, such that the net tissue architecture is a coordinated result of physical forces and a variety of cellular and extracellular factors.

VI. CELLS INVOLVED IN THE SYNTHESIS OF CORNEAL COLLAGEN

The embryonic chick cornea has served as a highly suitable model for proving the hypothesis (for review, see Hay, 1964) that epithelium as well as cells of the fibrocyte family produce collagen. In this section, we will describe the developing chick cornea with emphasis on the morphological and *in vitro* studies that establish the role of the epithelium in the production of ECM. Studies of collagen synthesis by the endothelium and by the fibroblasts will also be reviewed briefly and reference will be made to work on mammalian species as well.

Corneal development in the chick embryo proceeds in several distinct steps (Figs. 7–13), which are less definitive in the mammal where the stages presumably merge with each other (Coulombre, 1965; Hay and Revel, 1969; Pei and Rhodin, 1971; O'Rahilly, 1975; Ozanics *et al.*, 1976). After the lens detaches from overlying ectoderm, the ectoderm thickens and becomes the corneal epithelium (Fig. 7). The primary corneal stroma (Hay and Revel, 1969) now forms between the ectoderm and the lens capsule (Figs. 7 and 8). This stroma, initially 10 μm thick, consists of 25 layers of collagen fibrils which are orthogonally arranged and embedded in GAG. The basement membrane or basal lamina under the epithelium also contains GAG in a well-ordered pattern (Trelstad *et al.*, 1974), and presumably is rich in nonstriated collagen (Olsen *et al.*, 1973a) which contributes to its filamentous appearance (Fig. 8). The GAG particles are sensitive to digestion by chondroitinase (Trelstad *et al.*, 1974) and testicular hyaluronidase (Hay and Meier, 1974) and thus are likely to be chondroitin sulfate.

The evidence from the *in vivo* morphology of the corneal epithelium which suggests an epithelial origin of the collagenous primary stroma is as follows: the basal epithelial cells acquire the secretory organelles necessary for collagen synthesis (Fig. 9); the Golgi apparatus shifts to the basal cytoplasm as if to discharge its contents basally into the stroma (Hay and

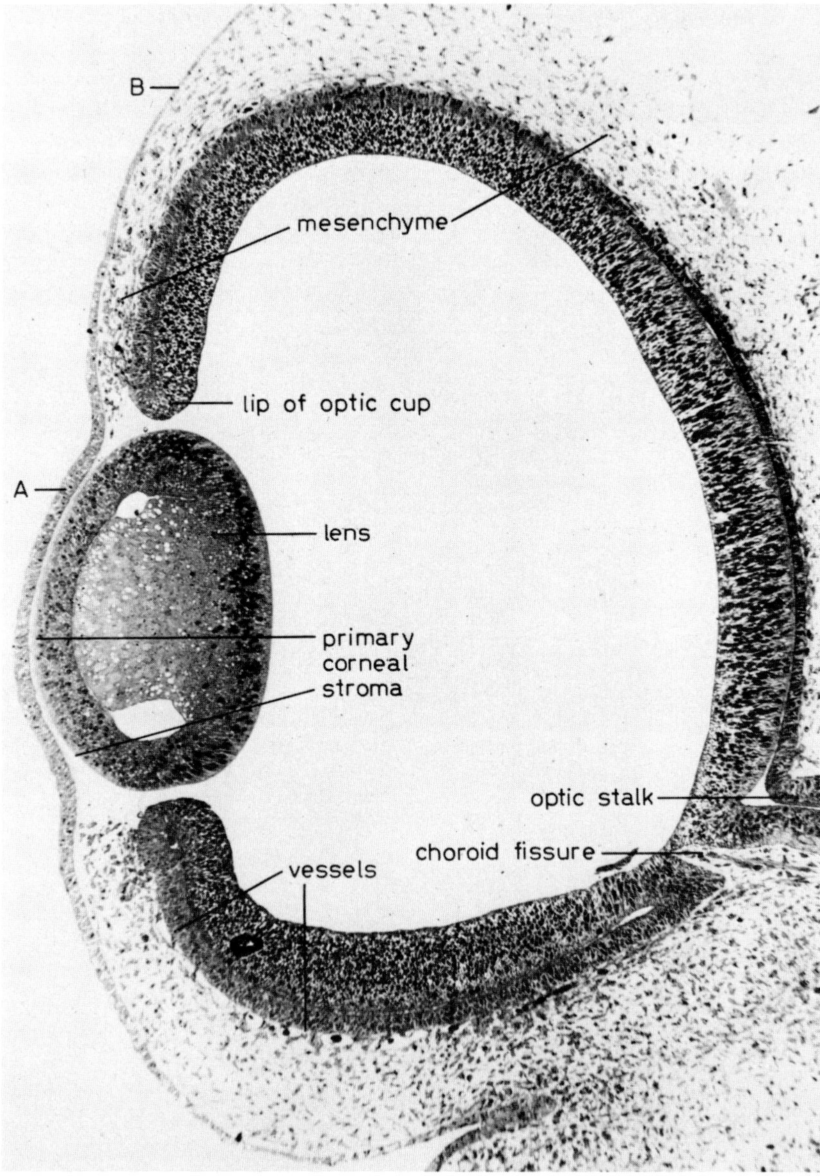

FIG. 7. Light micrograph of the embryonic chick eye at 4 days old (stage 22). The head ectoderm is 2 cells thick everywhere, but in the region of the cornea (A) it is taller than elsewhere (B), because here the basal cells have increased in height because of the accumulation of secretory organelles. The corneal epithelium secretes the primary corneal stroma (Dodson and Hay, 1971). × 7.7. (From Hay and Revel, 1969, by courtesy of S. Karger AG.)

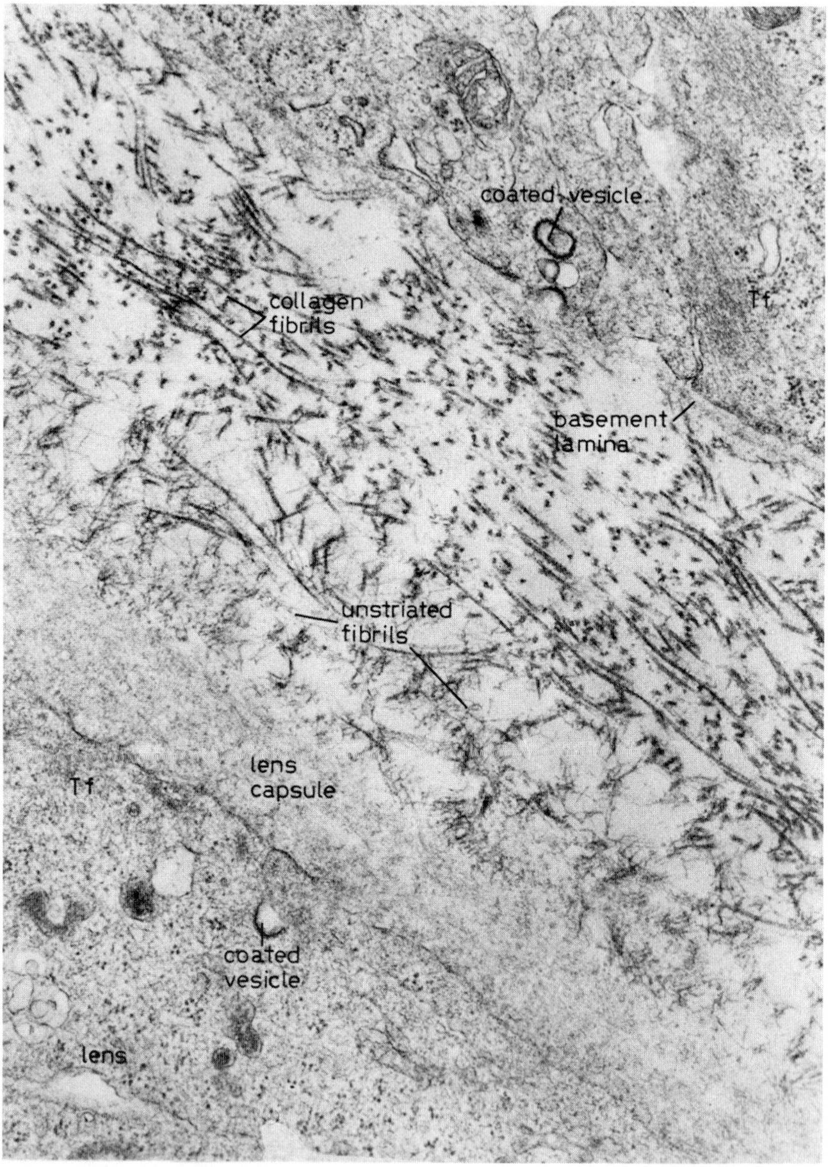

FIG. 8. Electron micrograph showing the primary corneal stroma at 4 days old (stage 22). The collagen fibrils at higher magnification show a typical, native, repeated banding. They are arranged in orthogonal layers, some of which are cut obliquely here because of the curvature of the cornea. (Tf) Tonofilaments. × 29,950. (From Hay and Revel, 1969, by courtesy of S. Karger AG.)

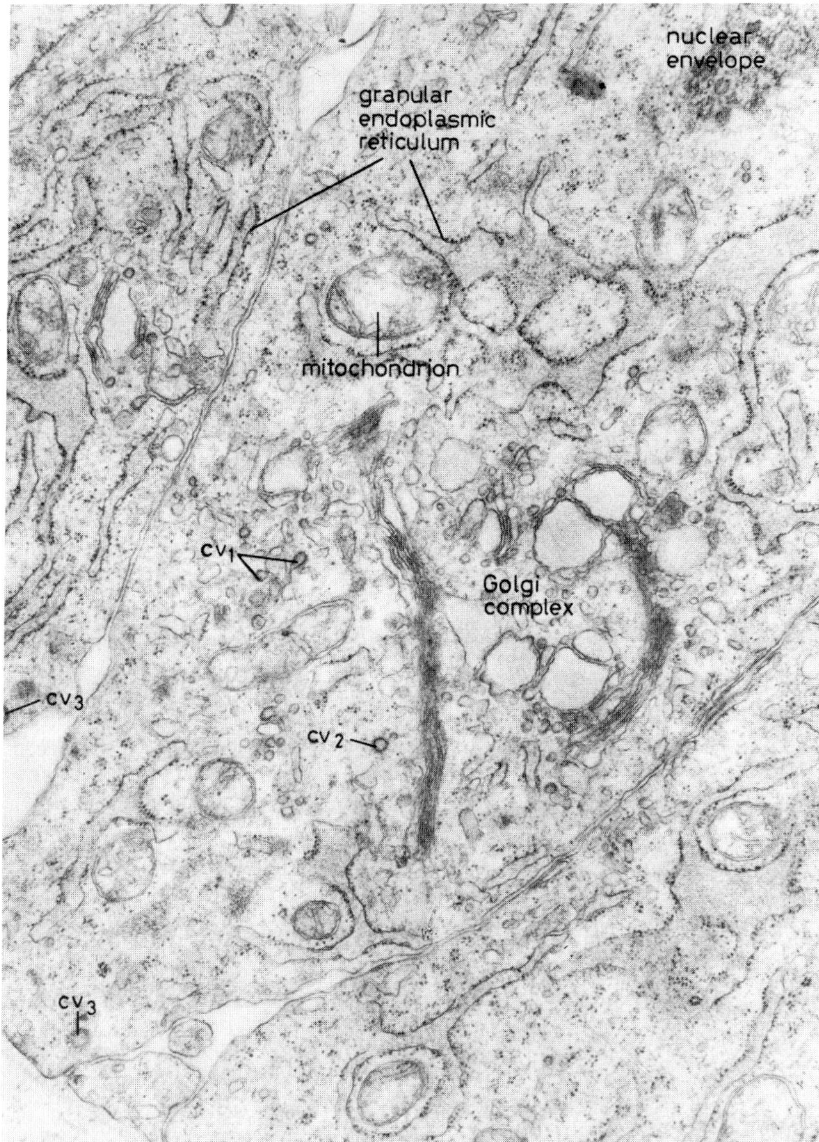

FIG. 9. Electron micrograph showing the well-developed Golgi complex and granular endoplasmic reticulum in a corneal epithelial cell from a 5½-day-old (stage 27) chick embryo. Coated vesicles are found attached to Golgi elements (cv_1), free in the cytoplasm (cv_2), and near the plasma membrane (cv_3). × 17,710. (From Hay and Revel, 1969, by courtesy of S. Karger AG.)

Revel, 1969; Trelstad, 1970); there are no mesenchymal cells in the area at stages 18–22 (Fig. 7) and those mesenchymal cells that are present around the periphery of the cornea do not possess well-developed secretory organelles. The lens epithelium produces a basement membrane (the lens capsule) and might also contribute collagen to the stroma at this time, but corneal fibroblasts do not make their appearance until much later and thus are not likely to be involved in primary stroma production.

In vitro studies of epithelial cell lines and isolated corneal epithelium reveal that epithelium is capable of incorporating ^{3}H-proline into hydroxyproline (Green and Goldberg, 1965; Goodfellow *et al.*, 1969; Conrad, 1970b; Gnadinger, 1972). Direct evidence that the corneal epithelium is producing native collagen fibrils was obtained by culturing isolated corneal epithelium (Dodson and Hay, 1971, 1974). The isolated epithelium is completely clean of basement membrane and other collagens at the beginning of the culture period. Cultured on a nonliving collagenous substratum, it secretes a facsimile of the corneal stroma within 24 hours *in vitro* (Fig. 10). The epithelium-derived fibrils have the native striated pattern typical of collagen, and a basement membrane forms in some regions (inset, Fig. 10). The new collagen is radioactive when ^{3}H-proline is added to the culture medium (Dodson and Hay, 1971; Hay and Dodson, 1973) and it can be collected and measured biochemically (Meier and Hay, 1974; 1975; Trelstad *et al.*, 1974; Linsenmayer *et al.*, 1977). Although a collagenous substratum enhances the production of collagen by the corneal epithelium, no detectable collagen from the substratum enters the epithelial cells or the new stroma (Hay, 1973; Hay and Meier, 1976).

The primary stroma secreted by the corneal epithelium plays an important role in the subsequent development of the cornea. In the next step of corneal differentiation in the chick embryo, mesenchymal cells destined to be the corneal endothelium use the collagenous primary stroma and lens capsule as substrata (Bard *et al.*, 1975; Nelson and Revel, 1975) to migrate between the stroma and the lens (Fig. 11). The endothelial cells attach to the posterior surface of the stroma, the anterior chamber forms, and then the stroma swells to a width of 50 μm (Hay and Revel, 1969). At the time of the swelling, the stroma becomes rich in hyaluronate (Toole and Trelstad, 1971), which seems to derive entirely from the endothelium; the corneal epithelium produces chondroitin sulfate, but not hyaluronate (Meier and Hay, 1973); and isolated corneal halves containing endothelium, but not epithelium, produce hyaluronate (Trelstad *et al.*, 1974). Moreover, the endothelial cells become linked by junctions at this time and may become active in water and ion transport. Therefore, the sudden swelling of the corneal stroma at stage 27 might be due to hydration induced by the high

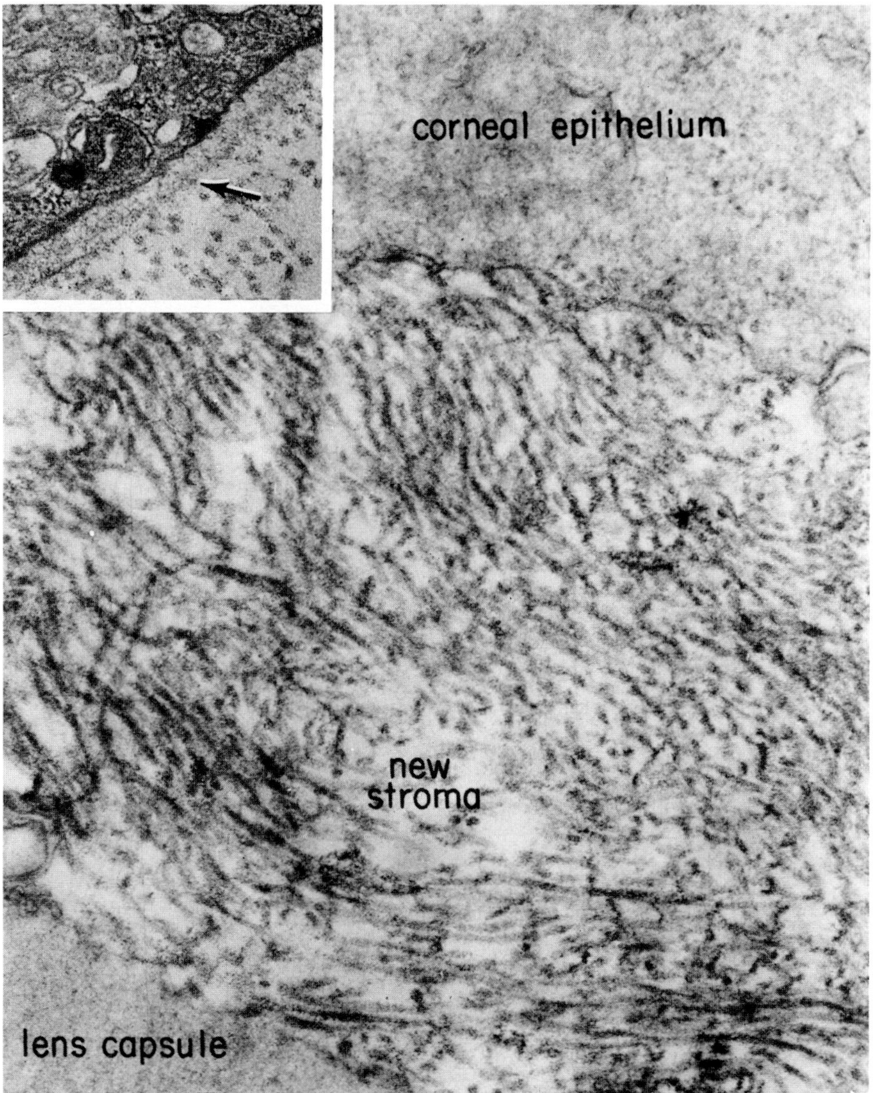

FIG. 10. Electron micrograph showing the new stroma produced by isolated corneal epithelium grown on NaOH-extracted lens capsule. The collagen fibrils of the epithelium-derived stroma are 25 nm wide and have a native striated pattern. × 70,000. (From Meier and Hay, 1974, by courtesy of Academic Press.) (Inset) The isolated corneal epithelium secretes a basement lamina over part of its basal surface (arrow). × 60,000. (From Hay and Dodson, 1973, by courtesy of the Rockefeller University Press.)

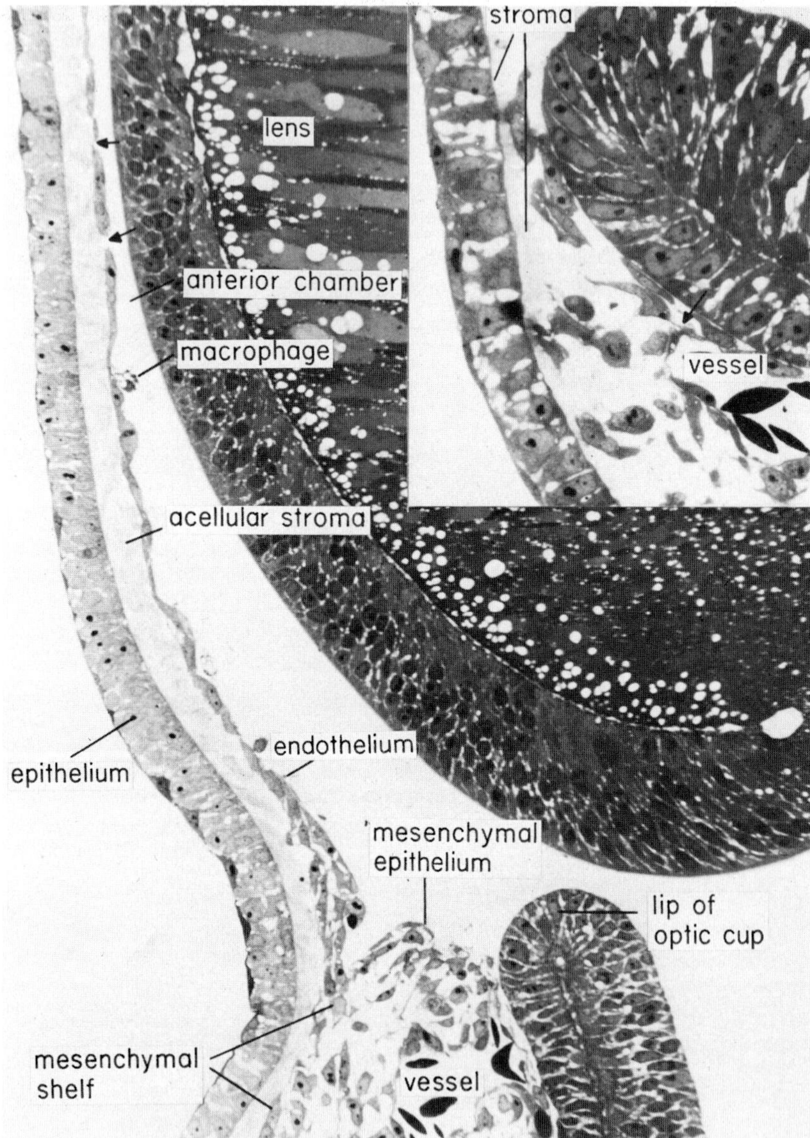

FIG. 11. Light micrograph of the embryonic chick cornea and lens at 5½ days old (stage 27). The endothelium has migrated between the primary (acellular) stroma to separate the cornea from the lens. At this stage, there are still gaps between the endothelial cells (arrows). The inset shows the angle of the eye at an earlier stage (stage 22). × 308. Inset, × 485. (From Hay and Revel, 1969, by courtesy of S. Karger AG.)

concentration of hyaluronate (Toole, 1976) and the possible transport activity of the endothelium (Hay and Revel, 1969).

The swelling of the corneal stroma is immediately followed by an invasion, at stages 27 and 28, of mesenchymal cells destined to become the corneal fibroblasts that will secrete the secondary (adult) stroma. That these cells use the layers of collagen fibers already in the primary stroma for their substratum is demonstrated by their relation to its complex orthogonal gridwork, by their behavior in defective primary stromas, and by their migration patterns *in vitro* (Trelstad and Coulombre, 1971; Coulombre and Coulombre, 1972, 1975; Bard and Hay, 1975). Between stages 35 and 40, the fibroblasts cease their migration and proliferation (Figs. 12 and 13a), hyaluronidase appears, and the hyaluronate content of

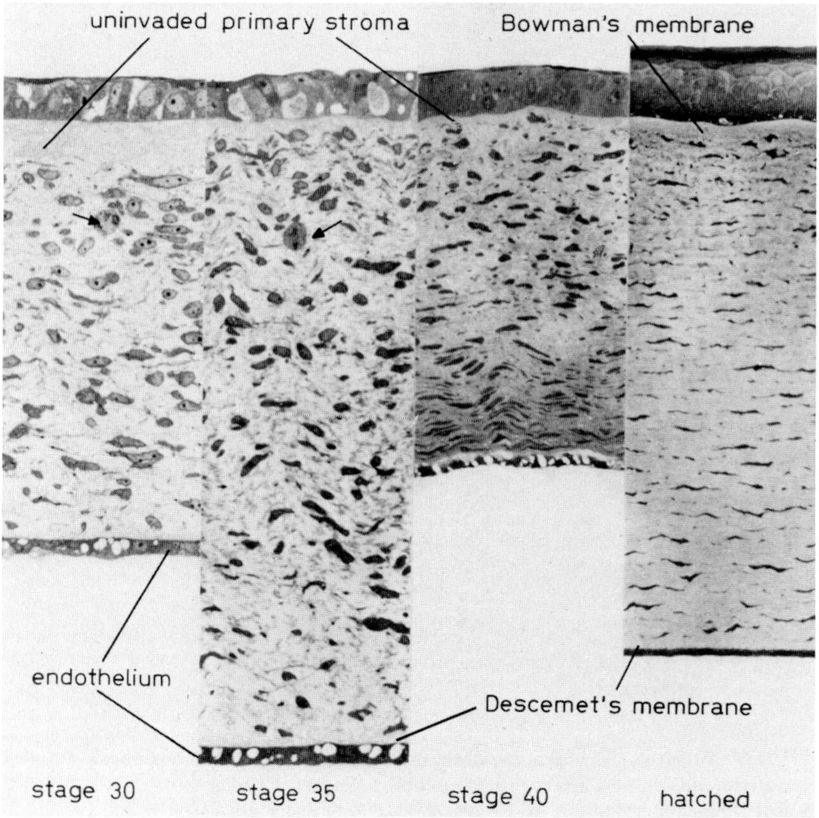

FIG. 12. Light micrographs of the embryonic chick cornea at 7, 9, and 14 days old, and after hatching. Between stage 35 (9 days old) and stage 40 (14 days old), the stroma begins to condense, loses both water and hyaluronate, and becomes rich in sulfated GAG. × 308. (From Hay and Revel, 1969, by courtesy of S. Karger AG.)

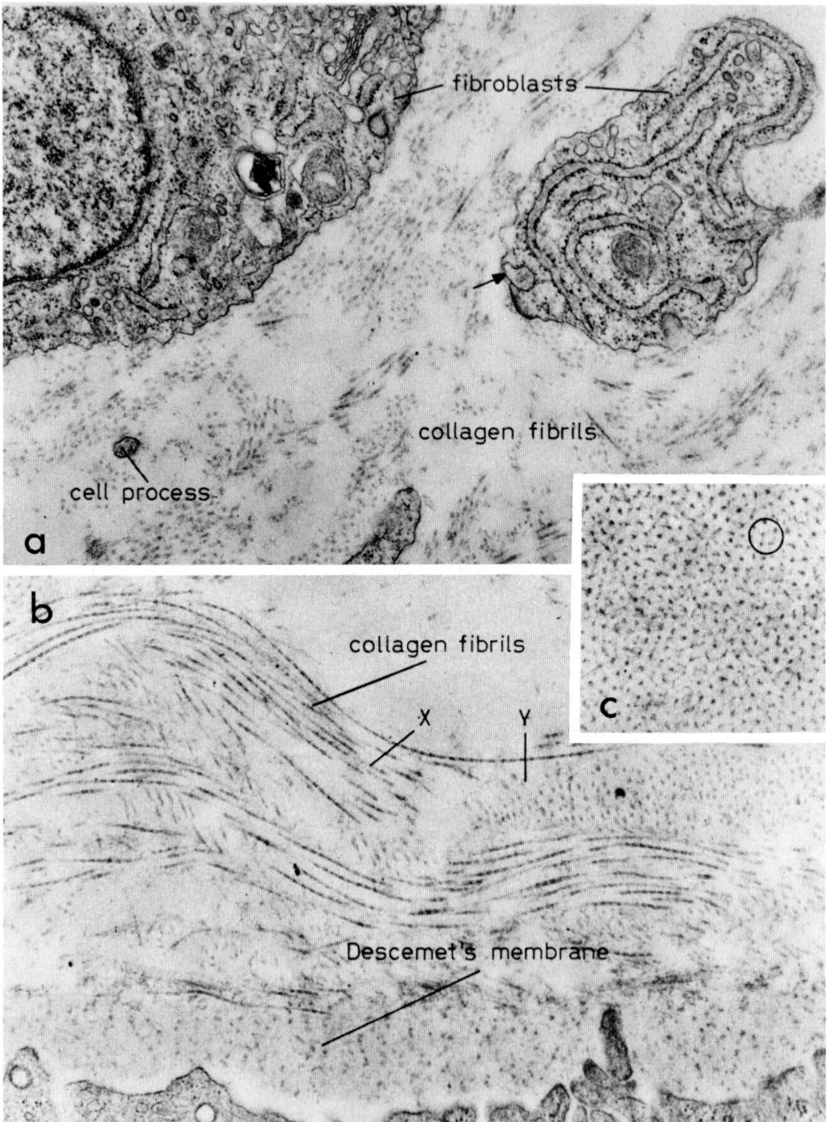

FIG. 13. Electron micrographs of cross sections of the anterior corneal stroma (a), posterior stroma (b), and a tangential section of Descemet's membrane (c) at stage 40 (14 days old). The collagen fibrils in the anterior stroma (a) are not yet as compacted as those in the posterior stroma (b). In Descemet's membrane at stage 40, 6–8 layers of small nodes are present (b), which in tangential section can be seen to be arranged in a hexagonal pattern (circle, c). (From Hay and Revel, 1969, by courtesy of S. Karger AG.)

the stroma diminishes (Toole and Trelstad, 1971). The stroma condenses (stage 40, Fig. 12) and chondroitin sulfate becomes the principal GAG produced (Toole and Trelstad, 1971). These remarkable changes, which are stimulated by the thyroid gland, are responsible for the dehydration and acquisition of the transparent form of the adult cornea (Coulombre and Coulombre, 1964; Coulombre, 1965).

From the beginning of fibroblast invasion, the fibroblasts produce keratan sulfate (Conrad and Dorfman, 1974; Hart, 1976), heparan sulfate (Conrad and Hart, 1975), chondroitin sulfate (Conrad, 1970a,b), and collagen (Hermann, 1961; Coleman *et al.*, 1965; Trelstad *et al.*, 1974). Keratan sulfate becomes more sulfated after day 14 and the ratios of 4-sulfated to 6-sulfated chondroitin sulfates increase during development, reaching a maximum on day 14 (Hart, 1976). The changing pattern of GAG synthesis may affect not only the water-binding capacity of the stroma (Toole, 1976), but also its refractive index and avascular nature (Hart, 1976). Changes in collagen types also occur during corneal morphogenesis (see Section VIII).

Interestingly, Descemet's membrane, a thick collagenous layer under the endothelium, first appears at stage 35 (9 days old), just prior to the beginning of the dehydration and corneal compaction described above (Fig. 12). By stage 40, Descemet's membrane consists of approximately 6 layers of dense nodes (each about 30 nm in diameter) embedded in an amorphous matrix (Fig. 13b). In a tangential section through one of the nodular layers, it can be seen that each node is surrounded by six others (Fig. 13c). This peculiar basement membrane is produced by the endothelium and may play a role in stromal condensation (see Hay and Revel, 1969, for review). Isolated mammalian corneal endothelium produces a basement-membrane type collagen *in vitro* (Kefalides *et al.*, 1976).

Bowman's membrane is a layer of compact, acellular, fibrillar connective tissue under the corneal epithelial basement membrane (Fig. 12). In the chick embryo, it represents the uninvaded primary corneal stroma (Hay and Revel, 1969) to which additional collagen and perhaps other antigens (Ikeda *et al.*, 1975) are added during development. It is undoubtedly produced by the epithelium in large part (Hay and Revel, 1969; Trelstad, 1970; Ikeda *et al.*, 1975). Both Descemet's membrane and Bowman's membrane contain type I and II collagens (see Section VIII), and they are probably rich in GAG.

VII. BIOCHEMICAL STUDIES OF COLLAGEN TYPES IN THE DEVELOPING AVIAN CORNEA

The initial biochemical studies that were done on the developing-chick primary corneal stroma suggested that type I collagen was the collagen composing the fibrils and that another collagen, containing $\alpha 1$ type chains, was also produced by the corneal epithelium (Trelstad *et al.*, 1974). It seemed reasonable at that time to believe that the second collagen would prove to be a typical basement-membrane collagen of the type IV class, but recent biochemical evidence (Linsenmayer *et al.*, 1977) suggests that this molecule is instead type II collagen.

Linsenmayer *et al.* (1977) cultured 5-day-old epithelium according to Dodson and Hay (1971) in the presence of ^{3}H-proline and β-aminopropionitrile, an agent that prevents crosslinking and thus facilitates extraction of the collagen (Levene and Gross, 1959). The newly synthesized, labeled collagen was extracted and purified in the presence of type I unlabeled carrier collagen from chick skin. CM-Cellulose chromatography (Fig. 14) suggested, as previously found by Trelstad *et al.* (1974), that two different types of collagen molecules had been produced and by fractional salt precipitation (Trelstad *et al.*, 1970; Mayne *et al.*, 1975) two

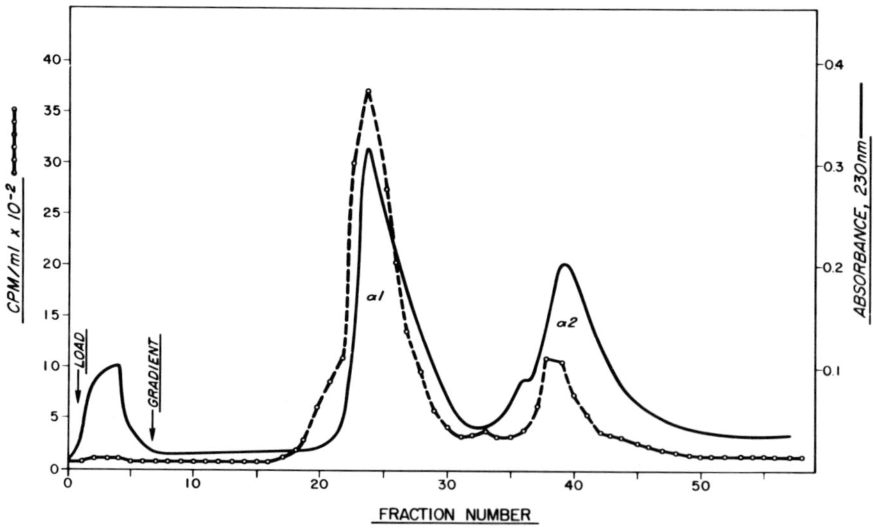

FIG. 14. **Chromatogram from a CM-cellulose column of the radioactive material purified from corneal epithelium cultures (–o—–o) plus unlabeled type I carrier collagen from chick skin (solid line). The solid line representing the optical density profile of the chick skin carrier shows the pattern characteristic for type I collagen with both $\alpha 1$ and $\alpha 2$ chain peaks in a relative ratio of 2:1. The collagen synthesized by the corneal epithelium culture also shows both $\alpha 1$ and $\alpha 2$ chain peaks, but here the relative ratio is about 4:1, suggesting the presence of two different molecules, one composed of both $\alpha 1$ and $\alpha 2$ type chains and another composed of $\alpha 1$ type chains only. (From Linsenmayer *et al.*, 1977, by courtesy of the National Academy of Science.)**

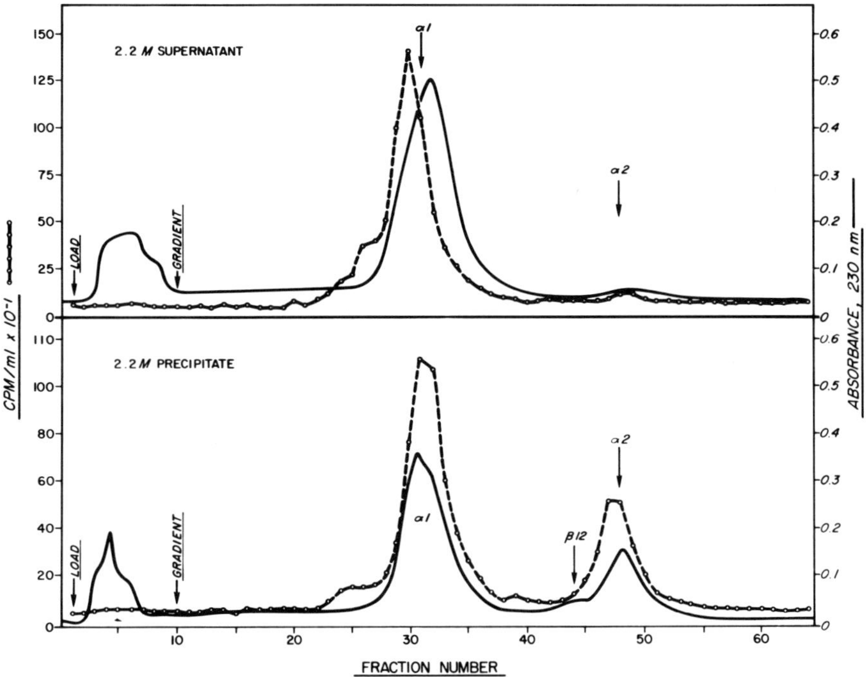

FIG. 15. Chromatograms from CM-cellulose columns of the radioactive material purified from corneal epithelium cultures (-○---○-) plus unlabeled type I and type II carrier collagens (solid lines) after fractional salt separation. At the top, material that remains in the supernatant at 2.2 *M* NaCl; at the bottom, material that precipitated at 2.2 *M* NaCl. The chromatograms show that the fractional salt separation procedure cleanly separated the type I carrier, which precipitated at 2.2 *M* NaCl, from the type II carrier which remained in the supernatant 2.2 *M* salt. Again, the type I carrier has both α1 and α2 chain peaks in the relative ratios of 2:1, whereas the type II carrier has only a single peak in the α1 chain region of the chromatogram. Note that the fractional salt separation procedure has also separated the corneal epithelium collagen into two components; one that coprecipitates with the type I carrier and has both α1 and α2 chains in a 2:1 ratio, and a second that stayed in the supernatant with the type II carrier and contains only α type chains. (From Linsenmayer *et al.*, 1977, by courtesy of the National Academy of Science.)

fractions were obtained, one composed of α1 type chains (top, Fig. 15) and the other composed of α1 and α2 chains in a ratio of 2:1 (bottom, Fig. 15).

To definitively identify a collagen type, it is necessary to obtain information on the amino acid sequence of the chains. For collagen, a convenient way to do this is to digest the molecules with cyanogen bromide (CNBr) that splits the component chains at the sites where methionine residues occur (see review by Miller, 1976). The two collagens synthesized by the corneal epithelium were digested with CNBr and the resulting peptides were separated by CM-cellulose chromatography (Fig. 16). The results indicate that one of the two different types of collagen produced by the corneal epithelium is similar, if not identical, to the type I molecule of skin, and the second is similar, if not identical, to the type II molecule of cartilage (Linsenmayer *et al.*, 1977). Further biochemical studies to define the presence or absence of a basement membrane collagen should complete the inventory of the collagens present during these critical early stages of chick corneal morphogenesis, since the immunofluorescent data indicate

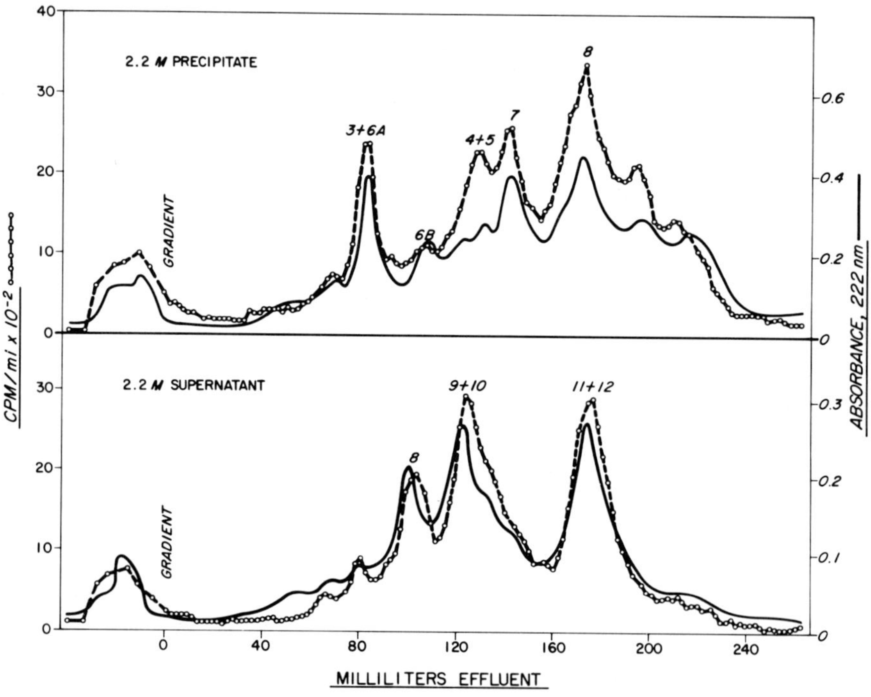

FIG. 16. Chromatograms from CM-cellulose columns of the peptides derived from CNBr cleavage of the fractionally salt separated, labeled collagens purified from corneal epithelial cultures (open circles) plus unlabeled carrier collagens (solid lines). At the top, material that precipitated at 2.2 *M* NaCl; at the bottom, material that remained in the supernatant at 2.2 *M* NaCl. The pattern of the major peptides produced by the carrier collagens and the corneal epithelium collagens is identical in both the 2.2 *M* precipitate which has the pattern of type I collagen, and the 2.2 *M* supernatant which has the pattern of type II collagen. The CNBr nomenclature for the labeled peptide peaks from the $\alpha 1$ type I and $\alpha 1$ type II chains is that of Mayne *et al.* (1975). The $\alpha 2$ peaks are not numbered (top right). (From Linsenmayer *et al.*, 1977, by courtesy of the National Academy of Science.)

the presence of type IV and absence of type III collagen in the primary corneal stroma.

VIII. IMMUNOHISTOCHEMICAL EVALUATION OF CORNEAL COLLAGEN HETEROGENEITY

The recent development of type-specific antibodies to collagen types I, II, III, and IV (Timpl, 1976; Timpl *et al.*, 1977) allow us now to extend the ultrastructural and biochemical studies described above because the specific localization of these various collagen types can be established in tissue sections. The effectiveness of the immunofluorescence technique using collagen type-specific antibodies has been demonstrated in numerous studies (for review, see Timpl, 1976; Timpl *et al.*, 1977). The distribution of collagen types in the developing chick eye has been studied recently (von der Mark *et al.*, 1977) using antibodies specific for chick type I and type II collagens (von der Mark *et al.*, 1976a,b), calf type III procollagen and collagen (Nowack *et al.*, 1976a), and against basement membrane from a mouse tumor (Timpl *et al.*, 1978), the latter two showing cross-reaction with the corresponding chick collagens (von der Mark and Timpl, in preparation).

The earliest primary corneal stroma examined (stage 19) seemed to contain type I collagen, but by stage 20 both types I and II collagen could be detected. These persist as a uniform mixture until stages 26 and 27 (Figs. 11 and 17), after which time the stroma is invaded by fibroblasts. The secondary stroma secreted by the fibroblasts seems to be almost entirely type I in composition, and type I continues to be found in the subepithelial and subendothelial layers (Fig. 18, left). Type II collagen persists in the subepithelial and subendothelial stroma (Fig. 18, right; Fig. 19; Fig. 20, middle) in fibrils which may be of mixed type I–type II origin (Fig. 21). Type IV collagen (basement-membrane collagen) is present under the epithelium (B, Fig. 22) and the endothelium, presumably in Descemet's membrane (En, Fig. 22), and is also found in the lens capsule and other basement membranes in the eyelid and sclera (von der Mark and Timpl, in preparation).

Remarkably, *no* type III collagen was found to be present in the chick corneal stroma at any stage (Fig. 20, right), although an unexplained reaction occurred over the early epithelium (von der Mark *et al.*, 1977). The adjacent eyelid and sclera are rich in type III collagen. This observation is unusual in that most tissues studied to date have both types I and III present in some admixture (Gay *et al.*, 1977). Thus, the highly uniform 25

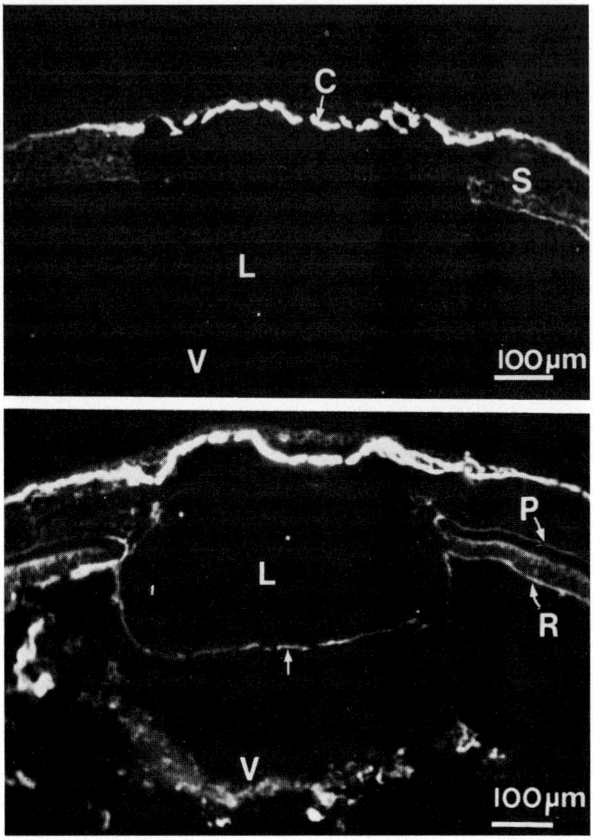

FIG. 17. Figures 17–20 are immunofluorescent light micrographs of chick corneas stained with antibodies to types I, II, and III collagen. Figure 17 shows that, at stage 26 (5 days old), the primary corneal stromal (C) stains for type I (top) and type II (bottom). Slight staining is also seen against type I in the sclera (S, top). Type II staining (bottom) is present in the vitreous body (V), and along the surfaces of the pigmented (P) and neural retina (R). The posterior surface of the lens (L) also stains for type II (arrow). × 80. (From von der Mark *et al.*, 1977, by courtesy of Academic Press.)

nm diameter fibrils in the secondary corneal stroma consist of a population of molecules made up almost entirely of type I collagen (Figs. 13 and 20), and these fibrils resemble the fibrils of mixed type I–type II derivation in the subepithelial stroma (Figs. 8 and 17). It will be interesting to find out whether or not the type I–type II collagens are especially suitable for the particular GAG–collagen interactions that occur in the cornea.

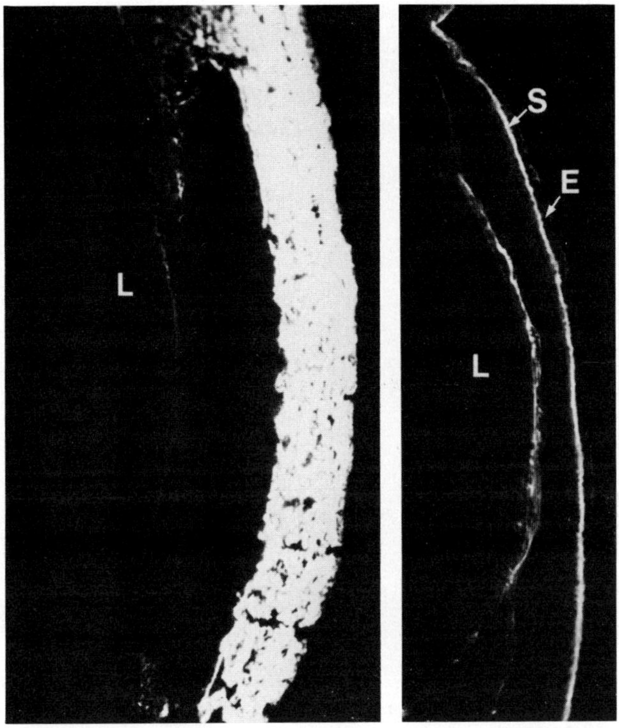

FIG. 18. By stage 30 (7 days old), the primary corneal stroma has been invaded by mesenchymal cells that synthesize type I collagen, and the entire corneal stroma stains intensely for type I (left). Type II collagen (right) is present at this stage principally in the uninvaded primary stroma (S) immediately beneath the corneal epithelium (E) and beneath the endothelium. The lens (L) and its anterior surface is unstained by anti-type II. × 80. (From von der Mark *et al.*, 1977, by courtesy of Academic Press.)

IX. SUMMARY

Collagens comprise a family of genetically distinct molecules, currently designated type I or $[\alpha 1(\mathrm{I})]_2\,\alpha 2$, type II or $[\alpha 1(\mathrm{II})]_3$, type III or $[\alpha 1(\mathrm{III})]_3$, and type IV or $[\alpha 1(\mathrm{IV})]_3$. The molecules are synthesized as procollagens which are converted by peptidases into the definitive collagen molecules. Immunochemical and autoradiographic studies indicate that procollagen is synthesized in the granular endoplasmic reticulum and together with GAG is secreted via the Golgi apparatus into the extracellular space, where final aggregation into fibrils and basement membranes takes place.

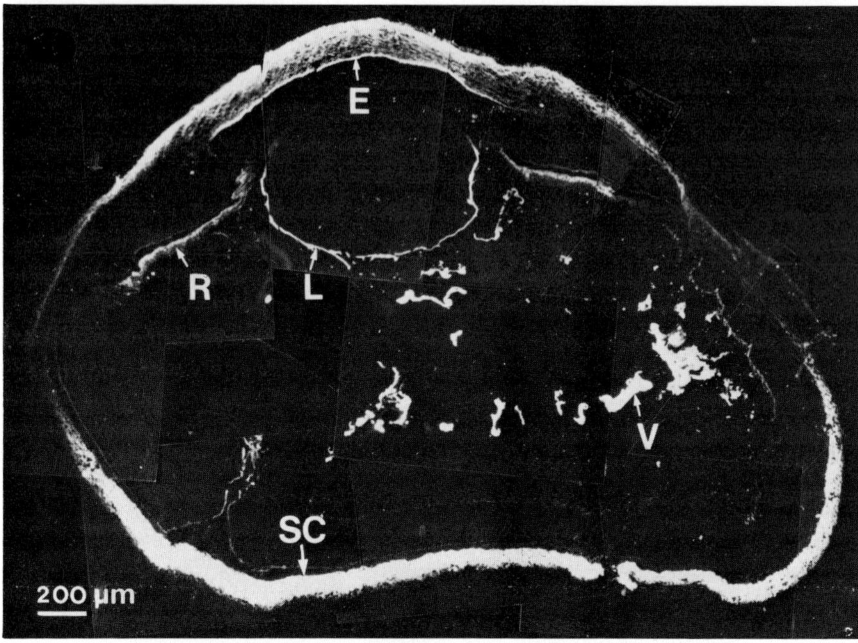

FIG. 19. A low-power view of a section through the whole eye at stage 32 (7½ days old) reveals type II collagen is especially concentrated in the subendothelial (E) and subepithelial stroma. Vitreous (V), scleral cartilage (SC), neural retina (R), and the posterior surface of the lens capsule (L) also contain type II collagen. × 40. (From von der Mark *et al.*, 1977, by courtesy of Academic Press.)

In the developing avian cornea, the corneal epithelium acquires highly developed endoplasmic reticulum and basally located Golgi complexes that enable it to synthesize and secrete the collagens that form the primary corneal stroma and the basement membrane. The orthogonally arranged fibrils are composed of both types I and II collagen, which can be detected biochemically and by immunochemistry. Cultures of isolated corneal epithelium clearly establish the origin of these collagens from the epithelium. By immunofluorescence, type IV collagen was located in the corneal stroma associated with the endothelium and epithelium, but type III collagen appears to be absent.

The primary corneal stroma which the epithelium produces between 3 and 4 days of development becomes separated from the lens on the fifth day by an endothelium that migrates in between the lens and stroma and comes to reside on the posterior surface of the stroma. Later, on the ninth day of development, the endothelium begins to produce the collagens

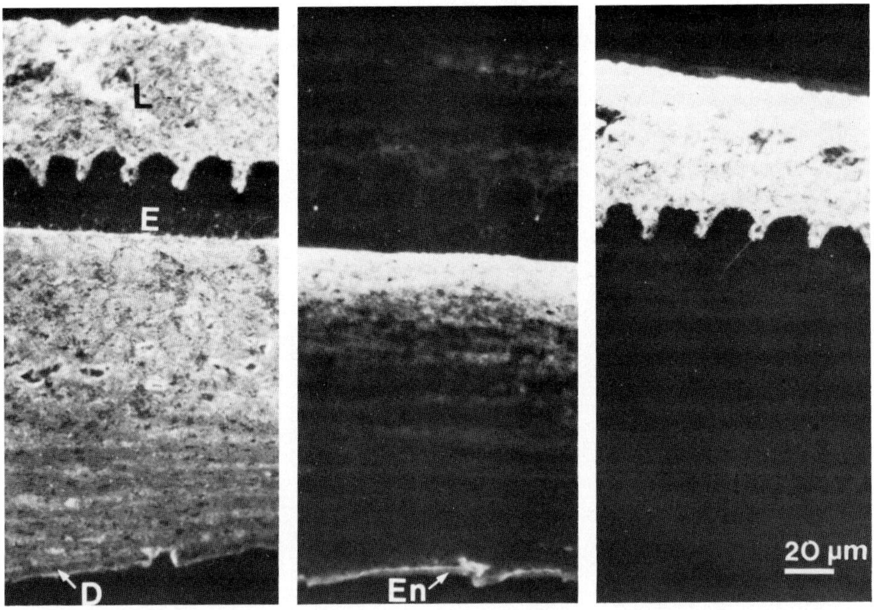

FIG. 20. With continued maturation the cornea at stage 42 (16 days old) continues to stain predominantly for type I collagen as does the eyelid (L, top left), while type II persists beneath the epithelium and endothelium (middle). Type III collagen stains the eyelid intensely, but there is strikingly no staining of the corneal stroma (right). (E) Epithelium, (En) endothelium, (D) Descemet's membrane. × 250. (From von der Mark *et al.*, 1977, by courtesy of Academic Press.)

(types II and IV) of Descemet's membrane. Meanwhile, on the seventh day, the primary stroma, now enriched with hyaluronate of endothelial origin, swells dramatically and is invaded by corneal fibroblasts. The corneal fibroblasts secrete the secondary corneal stroma (type I collagen) that lies between the posterior Descemet's membrane and the anterior remnant of the primary stroma that becomes Bowman's membrane. Type III collagen could not be detected in the corneal stroma at any age by immunochemistry, even though it is common in nearby embryonic connective tissues.

These studies on the development of corneal collagens, aside from their inherent interest to ophthalmologists and students of ocular development, are of general interest to developmental biologists studying morphogenetic mechanisms. The cornea is a relatively simple geometric structure with a limited number of macromolecular constituents and a relatively small number of cellular components. The progress made in studies of its fine

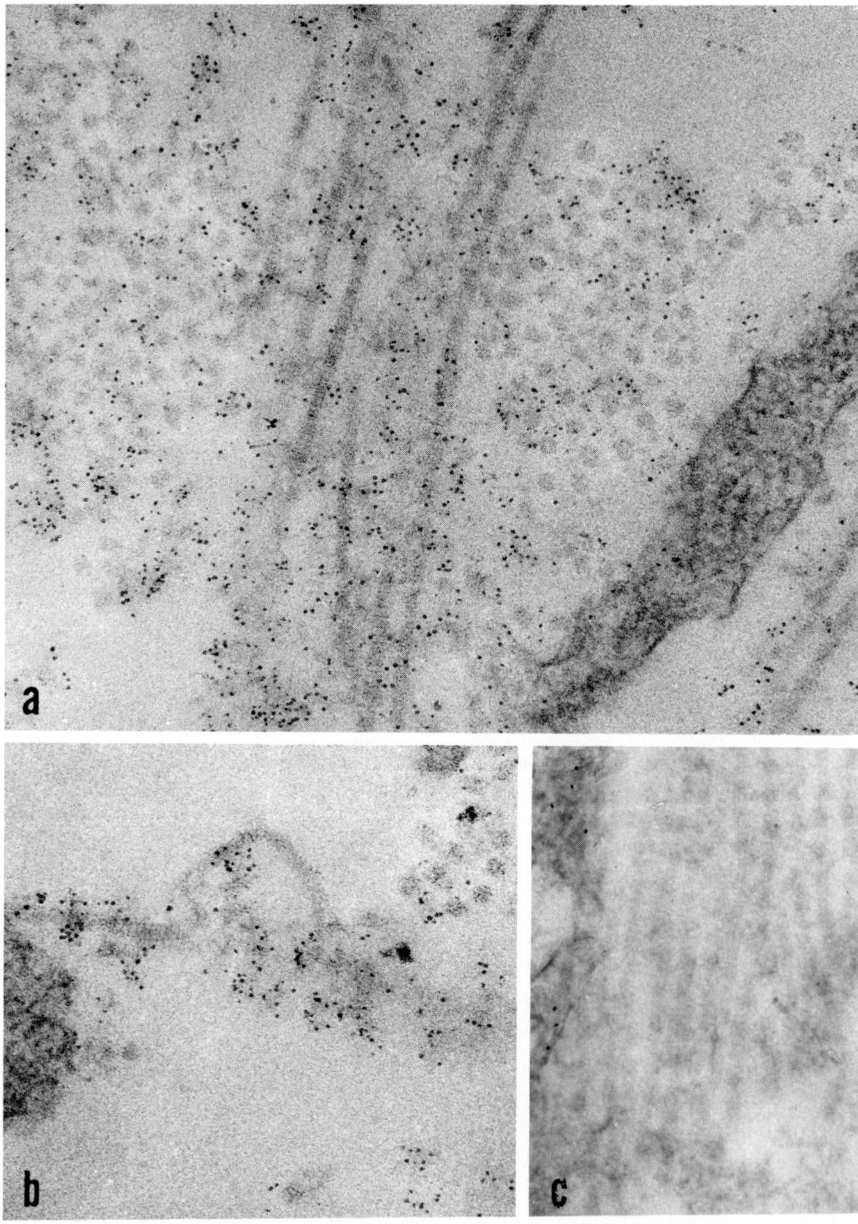

FIG. 21. Electron micrographs of 11-day-old (stage 37) chick corneas stained by the immunoferritin method show that both type I collagen (a) and type II collagen (b) are located in striated fibrils in the stroma. Small pieces of formaldehyde-fixed tissues were treated en bloc with rabbit anticollagen (a,b) or rabbit preimmune (unspecific) IgG (c), followed by an overnight wash. They were then exposed to ferritin-labeled goat antirabbit IgG, washed, postfixed, embedded, and sectioned. The fact that the ferritin label (black dots) is not uniform along the collagen fibril may be due in part to incomplete penetration of the antibody through the GAG-rich tissue en bloc. Collagen fibrils in the control (c) are not labeled. × 75,000. (Courtesy of E. D. Hay and K. von der Mark.)

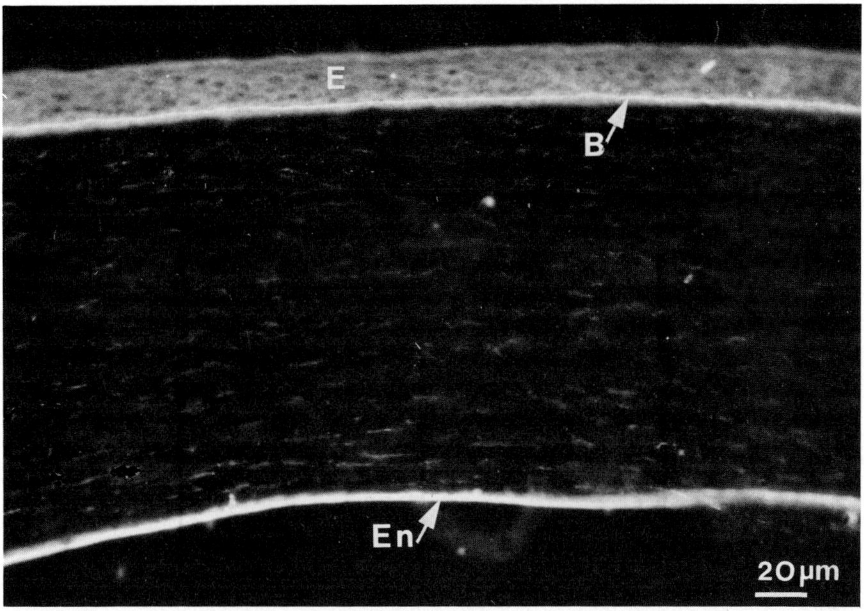

FIG. 22. Immunofluorescence staining of a section of a stage 45 (19-day-old) chick cornea with antibodies to basement membrane collagen. This collagen type (type IV) is located in Bowman's membrane (B), probably in the basement membrane, and is associated with the endothelium (En), probably in Descemet's membrane. The weak staining of the epithelium (E) and fibroblasts is probably unspecific. × 250. (Courtesy of K. von der Mark and R. Timpl.)

structure, its biochemistry, and its cellular behavior suggest that continued investigation of this tissue will produce substantial information of relevance to our understanding not only of corneal biology, but also of extracellular matrix biogenesis.

Acknowledgments

Dr. Linsenmayer is a recipient of an N.I.H. Research Career Development Award, 1–K04 AM0031. Dr. Trelstad is a recipient of an American Cancer Society Faculty Award, PR–107.

References

Bard, J. B. L., and Hay, E. D. (1975). *J. Cell Biol.* **67**, 400–418.

Bard, J. B. L., Hay, E. D., and Meller, S. M. (1975). *Dev. Biol.* **42**, 334–361.

Byers, P., McKenney, K., Lichtenstein, J. R., and Martin, G. R. (1974). *Biochemistry* **13**, 5243–5247.

Byers, P. H., Click, E. M., Harper, E., and Bornstein, P. (1975). *Proc. Natl. Acad. Sci. U.S.A.* **72**, 3009–3013.

Bornstein, P. (1974). *Annu. Rev. Biochem.* **43**, 567–603.

Bornstein, P., Ehrlich, H. P., and Wyke, A. W. (1972). *Science* **175**, 544–546.

Burgeson, R. E., El Adli, F. A.,Kaitila, I. I., and Hollister, D. W. (1976). *Proc. Natl. Acad. Sci. U.S.A.* **73**, 2579–2583.

Chung, E., and Miller, E. J. (1974). *Science* **183**, 1200–1201.

Chung, E., Rhodes, K., and Miller, E. J. (1976). *Biochem. Biophys. Res. Commun.* **71**, 1167–1174.

Church, R. L., Tanzier, M. L., and Lapiere, C. M. (1973). *Nature (London), New Biol.* **244**, 188–190.

Church, R. L., Yaeger, J. A., and Tanzer, M. L. (1974). *J. Mol. Biol.* **86**, 785–799.

Cintron, C. (1974). *Biochem. Biophys. Res. Commun.* **60**, 288–294.

Coleman, J. R., Herrmann, H., and Bess, B. (1965). *J. Cell Biol.* **25**, 69–78.

Conrad, G. W. (1970a). *Dev. Biol.* **21**, 292–317.

Conrad, G. W. (1970b). *Dev. Biol.* **21**, 611–635.

Conrad, G. W., and Dorfman, A. (1974). *Exp. Eye Res.* **18**, 421–433.

Conrad, G. W., and Hart, G. W. (1975). *Dev. Biol.* **44**, 253–269.

Coulombre, A. J. (1965). *Adv. Morphog.* **4**, 81–109.

Coulombre, A. J., and Coulombre, J. L. (1964). *Exp. Eye Res.* **3**, 105–114.

Coulombre, A. J., and Coulombre, J. L. (1972). *Dev. Biol.* **28**, 183–190.

Coulombre, J. L., and Coulombre, A. J. (1975). *Dev. Biol.* **45**, 291–303.

Dehm, P., and Prockop, D. J. (1971). *Biochim. Biophys. Acta* **240**, 358–369.

Dehm, P., and Prockop, D. J. (1972). *Biochim. Biophys. Acta* **262**, 375–382.

Dehm, P., and Prockop, D. J. (1973). *Eur. J. Biochem.* **35**, 159–166.

Dehm, P., Olsen, B. R., and Prockop, D. J. (1974). *Eur. J. Biochem.* **46**, 107–116.

Dodson, J. W., and Hay, E. D. (1971). *Exp. Cell Res.* **65**, 215–220.

Dodson, J. W., and Hay, E. D. (1974). *J. Exp. Zool.* **189**, 51–72.

Ehrlich, H. P., and Bornstein, P. (1972). *Biochem. Biophys. Res. Commun.* **46**, 1750–1754.

Ehrlich, H. P., Ross, R., and Bornstein, P. (1974). *J. Cell Biol.* **62**, 390–405.

Epstein, E. H., Jr. (1974). *J. Biol. Chem.* **249**, 3225–3231.

Fessler, L. I., Burgeson, R. E., Morris, N. P., and Fessler, J. H. (1973). *Proc. Natl. Acad. Sci. U.S.A.* **70**, 2993–2996.

Fessler, L. I., Morris, N. P., and Fessler, J. H. (1975). *Proc. Natl. Acad. Sci. U.S.A.* **72**, 4905–4909.

Fietzek, P. P., and Kühn, K. (1976). *Int. Rev. Connect. Tissue Res.* **7**, 1–60.

Furthmayr, H., Timpl, R., Stark, M., Lapiere, C. M., and Kühn, K. (1972). *FEBS Lett.* **28**, 247.

Furthmayr, H., and Timpl, R. (1976). *Int. Rev. Connect. Tissue Res.* **7**, 61–99.

Gay, S., Martin, G. R., Muller, P. K., Timpl, R., and Kuhn, K. (1977). *Proc. Natl. Acad. Sci. U.S.A.* **73**, 4037–4040.

Gnadinger, M. C. (1972). *Ophthalmic Res.* **3**, 198–206.

Goldberg, B. E. Epstein, E. H., Jr., and Sherr, C. J. (1972). *Proc. Natl. Acad. Sci. U.S.A.* **69**, 3655–3659.

Goodfellow, R. I., Revel, J. P., and Hay, E. D. (1969). *Anat. Rec.* **163**, 191.

Grant, M. E., and Prockop, D. J. (1972). *N. Engl. J. Med.* **285**, 194–199, 242–248, 291–300.

Green, H., and Goldberg, B. (1965). *Proc. Natl. Acad. Sci. U.S.A.* **73**, 1360–1362.

Gross, J. (1956). *J. Biophys. Biochem. Cytol.* **2**, 261–273.

Gross, J. (1973). *Harvey Lect. Ser.* **68**, 351–432.

Gross, J., and Kirk, D. (1958). *J. Biol. Chem.* **233**, 355–360.

Hart, G. W. (1976). *J. Biol. Chem.* **251**, 6513–6521.

Hart, R. W., and Farrell, R. A. (1969). *J. Opt. Soc. Am.* **59**, 766–774.

Hay, E. D. (1964). *In* "The Epidermis" (W. Montagna and W. C. Lobitz, eds.), pp. 97–116. Academic Press, New York.
Hay, E. D. (1973). *Am. Zool.* **13,** 1085–1107.
Hay, E. D., and Dodson, J. W. (1973). *J. Cell Biol.* **57,** 190–213.
Hay, E. D., and Meier, S. (1974). *J. Cell Biol.* **62,** 889–898.
Hay, E. D., and Meier, S. (1976). *Dev. Biol.* **52,** 141–157.
Hay, E. D., and Revel, J. P. (1963). Dev. Biol. **7,** 152–168.
Hay, E. D., and Revel, J. P. (1969). *In* "Monographs in Developmental Biology" (A. Wolsky and P. S. Chen, eds.), Vol. 1, pp. 1–144. Karger, Basel.
Hermann, H. (1961). *In* "The Structure of the Eye" (G. K. Smelser, ed.), pp. 421–433. Academic Press, New York.
Ikeda, A., Maisel, H., and Waggoner, D. (1975). *J. Embryol. Exp. Morphol.* **33,** 279–290.
Jamieson, G. A., Urban, C. L., and Barber, A. J. (1971). *Nature (London), New Biol.* **234,** 5–7.
Kang. A. H., Beachy, E. H., and Katzman, R. L. (1974). *J. Biol. Chem.* **249,** 1054–1059.
Katzman, R., Kang. A. H., and Dixit, S. N. (1974). *Biochim. Biophys. Acta* **336,** 367–369.
Kefalides, N. A. (1973). *Int. Rev. Connect. Tissue Res.* **6,** 63–104.
Kefalides, N. A. (1975). *J. Invest. Dermatol.* **65,** 85–92.
Kefalides, N. A., Cameron, J. D., Tomichek, E. A., and Yanoff, M. (1976). *J. Biol. Chem.* **251,** 730–733.
Lapiere, C. M., Lanaers, A., and Kohn, L. D. (1971). *Proc. Natl. Acad. Sci. U.S.A.* **68,** 3054–3058.
Layman, D. L., McGoodwin, E. B., and Martin, G. R. (1971). *Proc. Natl. Acad. Sci. U.S.A.* **68,** 454–458.
Levene, C. I., and Gross, J. (1959). *J. Exp. Med.* **110,** 771–790.
Lichtenstein, J. R., Byers, P. H., Smith, B. O., and Martin, G. R. (1975). *Biochemistry* **14,** 1589–1594.
Linsenmayer, T. F. (1974). *Dev. Biol.* **40,** 372–377.
Linsenmayer, T. F., Trelstad, R. L., Toole, B. P., and Gross, J. (1973a). *Biochem. Biophys. Res. Commun.* **52,** 870–876.
Linsenmayer, T. F., Trelstad, R. L., and Gross, J. (1973b). *Biochem. Biophys. Res. Commun.* **53,** 39–45.
Linsenmayer, T. F., Toole, B. P., and Trelstad, R. L. (1973c). *Dev. Biol.* **35,** 232–239.
Linsenmayer, T. F., Smith, G. N., and Hay, E. D. (1977). *Proc. Natl. Acad. Sci. U.S.A.* **74,** 39–43.
Martin, G. R., Byers, P. H., and Piez, K. A. (1975). *Adv. Enzymol. Relat. Areas Mol. Biol.* **42,** 167–191.
Mayne, R., Vail, M. S., and Miller, E. J. (1975). *Proc. Natl. Acad. Sci. U.S.A.* **72,** 4511–4515.
Meier, S., and Hay, E. D. (1973). *Dev. Biol.* **35,** 318–331.
Meier, S., and Hay, E. D. (1974). *Dev. Biol.* **38,** 249–270.
Meier, S., and Hay, E. D. (1975). *J. Cell Biol.* **66,** 275–291.
Miller, A., and Parry, D. A. G. (1973). *J. Mol. Biol.* **75,** 441–447.
Miller, E. J. (1976). *Mol. Cell. Biochem.* **13,** 165–192.
Miller, E. J., and Matthews, M. B. (1974). *Biochem. Biophys. Res. Commun.* **60,** 424–430.
Miller, E. J., and Matukas, V. J. (1969). *Proc. Natl. Acad. Sci. U.S.A.* **64,** 1264–1268.
Miller, E. J., and Matukas, V. J. (1974). *Fed. Proc., Fed. Am. Soc. Exp. Biol.* **33,** 1197–1204.
Miller, E. J., Martin, G. R., Piez, K. A., and Powers, M. J. (1967). *J. Biol. Chem.* **242,** 5481–5489.

Miller, E. J., Epstein, E. H., Jr., and Piez, K. (1971). *Biochem. Biophys. Res. Commun.* **42,** 1024–1029.
Minor, R. R., Clark, C. C., Strause, E. L., Koszalka, T. R. Brent, R. L., and Kefalides, N. A. (1976). *J. Biol. Chem.* **251,** 1789–1794.
Monson, J. M., Click, E. M., and Bornstein, P. (1975). *Biochemistry* **14,** 4088–4092.
Nelson, G. A., and Revel, J. P. (1975). *Dev. Biol.* **42,** 315–333.
Newsome, D. A. Linsenmayer, T. F., and Trelstad, R. L. (1976). *J. Cell Biol.* **71,** 59–67.
Nist, C., von der Mark, K., Hay, E. D., Olsen, B. R., Bornstein, P., Ross, R., and Dehm, P. (1975). *J. Cell Biol.* **65,** 75–87.
Nowack, H., Gay, S., Wick, G., Becker, V., and Timpl, R. (1976a) *J. Immunol. Methods* **12,** 117–124.
Nowack. H., Olsen, B. R., and Timpl, R. (1976b). *Eur. J. Biochem.* **70,** 205–216.
Olsen, B. R., and Prockop, D. J. (1974). *Proc. Natl. Acad. Sci. U.S.A.* **71,** 2033–2037.
Olsen, B. R., Alper, R., and Kelfalides, N. A. (1973a). *Eur. J. Biochem.* **38,** 220–228.
Olsen, B. R., Berg. R. A., Kishida, Y., and Prockop, D. J. (1973b). *Science* **182,** 825–827.
Olsen, B. R., Berg, R. A., Kishida, Y., and Prockop, D. J. (1975). *J. Cell Biol.* **64,** 340–355.
O'Rahilly, R. (1975). *Exp. Eye Res.* **21,** 93–112.
Ozanicks, V., Rayborn, M., and Sagun, D. (1976). *Exp. Eye Res.* **22,** 305–327.
Pei, Y. F., and Rhodin, J. A. B. (1971). *Invest. Ophthalmol.* **10,** 811–825.
Piez, K. A. Eigner, E. A., and Lewis, M. S. (1963). *Biochemistry* **2,** 58–66.
Prockop, D. J., Berg, R. A., Kivirikko, K. I., and Uitto, J. (1976). *In* "Biochemistry of Collagen" (G. N. Ramachandran and A. H. Reddi, eds.), pp. 163–273. Plenum, N.Y.
Revel. J. P. (1964). *J. Microsc. (Paris)* **3,** 535–544.
Revel, J. P. (1970). *In* "Chemistry and Molecular Biology of the Extracellular Matrix" (E. A. Balazs, ed.), pp. 1485–1502. Academic Press, New York.
Revel, J. P., and Hay, E. D. (1963). *Z. Zellforsch. Mikrosk. Anat.* **61,** 110–144.
Ross, R., and Benditt, E. P. (1965). *J. Cell Biol.* **27,** 83–106.
Salpeter, M. M. (1968). *J. Morphol.* **124,** 387–422.
Schofield, J. D., and Prockop, D. J. (1973). *Clin. Orthop. Relat. Res.* **97,** 175–195.
Schofield, J. D., Freeman, I. L., and Jackson, D. S. (1971). *Biochem. J.* **124,** 467–473.
Sherr, C. J., and Goldberg, B. (1973). *Science* **180,** 1190–1192.
Smith, B. D., Byers, P. H., and Martin, G. R. (1972). *Proc. Natl. Acad. Sci. U.S.A.* **69,** 3260–3262.
Smith, G. N., Jr., Linsenmayer, T. F., and Newsome, D. A. (1976). *Proc. Natl. Acad. Sci. U.S.A.* **73,** 4420–4423.
Swann, D. A., Constable, S. J., and Harper, E. (1972). *Invest. Ophthalmol.* **11,** 735–738.
Tanzer, M. L., Church, R. L., Yaeger, J. A., Wampler, D. E., and Park, E. (1974). *Proc. Natl. Acad. Sci. U.S.A.* **71,** 3009–3013.
Timpl, R. (1976). *In* "Biochemistry of Collagen" (G. N. Ramachandran and A. H. Reddi, eds.), pp. 319–375. Plenum, New York.
Timpl, R., Wick, G., Furthmayr, H., Lapiere, C. M., and Kuhn, K. (1973). *Eur. J. Biochem.* **32,** 584–591.
Timpl, R., Martin, G. R., Bruckner, P., Wick, G., and Wiedemann, H. (1978). *Eur. J. Biochem.* **84,** 43–52.
Timpl, R., Wick, G., and Gay, S. (1977). *J. Immunol. Meth.* **18,** 165–182.
Toole, B. P. (1976). *In* "Neuronal Recognition" (S. Barondes, ed.), pp. 275–329. Plenum, New York.
Toole, B. P., and Lowther, D. A. (1968a). *Biochem. J.* **109,** 857–866.
Toole, B. P., and Lowther, D. A. (1968b). *Arch. Biochem. Biophys.* **128,** 567–578.
Toole, B. P., and Trelstad, R. L. (1971). *Dev. Biol.* **26,** 28–35.

Toole, B. P., Kang, A. H., Trelstad, R. L., and Gross, J. (1972). *Biochem. J.* **127,** 715–720.
Trelstad, R. L. (1970). *J. Cell Biol.* **45,** 34–42.
Trelstad, R. L. (1971). *J. Cell Biol.* **48,** 689–694.
Trelstad, R. L. (1973). *J. Histochem. Cytochem.* **21,** 521–528.
Trelstad, R. L. (1974). *Biochem. Biophys. Res. Commun.* **57,** 717–725.
Trelstad, R. L., and Coulombre, A. J. (1971). *J. Cell Biol.* **50,** 840–858.
Trelstad, R. L., and Kang, A. H. (1974). *Exp. Eye Res.* **18,** 395–406.
Trelstad, R. L., and Lawley, K. R. (1977). *Biochem. Biophys. Res. Commun.* **76,** 376–384.
Trelstad, R. L., Kang, A. H., Igarashi, S., and Gross, J. (1970). *Biochemistry* **9,** 4993–4998.
Trelstad, R. L., Kang, A. H., Toole, B. P., and Gross, J. (1972). *J. Biol. Chem.* **247,** 6469–6473.
Trelstad, R. L., Hayashi, K., and Toole, B. P. (1974). *J. Cell Biol.* **62,** 815–830.
Trelstad, R. L., Hayashi, K., and Gross, J. (1976). *Proc. Natl. Acad. Sci. U.S.A.* **73,** 4027–4031.
Uitto, J., Lichtenstein, J. R., and Bauer, E. A. (1976). *Biochemistry* **15,** 4935–4942.
von der Mark, K., and Bornstein, P. (1973). *J. Biol. Chem.* **248,** 2285–2289.
von der Mark, K., Click, E. M., and Bornstein, P. (1973). *Arch. Biochem. Biophys.* **156,** 356–364.
von der Mark, H., von der Mark, K., and Gay, S. (1976a). *Dev. Biol.* **48,** 237–249.
von der Mark, K., von der Mark, H., and Gay, S. (1976b). *Dev. Biol.* **53,** 153–170.
von der Mark, K., von der Mark, H., Timpl, R., and Trelstad, R. L. (1977). *Dev. Biol.* **59,** 75–85.
Weinstock, M., and Leblond, C. P. (1974). *J. Cell Biol.* **60,** 92–127.
Wood, G. C. (1964). *Int. Rev. Connect. Tissue Res.* **2,** 1–31.
Young, R. W. (1973). *J. Cell Biol.* **57,** 175–189.

The Electrophysiology of the Crystalline Lens

JAMES L. RAE[1]

I.	Introduction	37
II.	Electrophysiologic Morphology	38
	A. Capsule	38
	B. Epithelium	39
	C. Lens Fibers	40
	D. Cell-to-Cell Junctions and Interdigitations	41
	E. Deep Fiber Membranes	49
III.	Extracellular Space	49
	A. Electron Microscope Studies	49
	B. Isotope Studies	50
	C. Summary	53
IV.	Electrophysiology	54
	A. Fiber Potential vs. Outside Layer Potential	54
	B. The Lens as an Epithelial Membrane	61
	C. Resistance Measurements	66
	D. Ion Distribution and Movement	71
	E. Origin of the Potential Difference	83
V.	Summary	84
	References	85

I. INTRODUCTION

The work presented here is intended to be a specific review of the electrophysiology of the crystalline lens. Even though a significant amount of lens literature is cited, no attempt is made to be exhaustive. In light of recent information, it is clear that one must understand the morphology of the lens if one wishes to appreciate its electrophysiology. Therefore, a section on electrophysiologic morphology is included. The content of the

[1] Departments of Ophthalmology and Physiology and Biophysics, University of Texas Medical Branch, Galveston, Texas.

ISBN 0-12-153001-9

review revolves heavily around morphologic, potential difference, electrical resistance, and ion-flux measurements. Although it is certain that energy metabolism, sugar transport, amino acid transport, and H_2O metabolism are important to lens electrophysiology at some level, their relationships are not sufficiently direct to be covered here. Also, the areas of natural cataracts, experimental cataracts, and drug effects are omitted.

It is assumed that the reader is familiar with the basic concepts of bioelectricity including the significance of ionic fluxes, Nernst potentials, and the Goldman–Hodgkin–Katz equation, thus obviating the development of such topics here. Throughout the review, attempts are made to point to areas where, in the author's opinion, gaps in knowledge seem to exist. Appropriate literature is cited so that the reader may consider each topic more completely.

II. ELECTROPHYSIOLOGIC MORPHOLOGY

Mammalian, fish, and amphibian lenses have some common morphologic features which are presumably important to their electrophysiology. This discussion is limited primarily to these lenses since they have been the subjects of the majority of lens electrophysiologic investigations. Also, a considerable number of recent electron microscopic studies exist for these lenses (Wanko and Gavin, 1958, 1959, 1961; Cohen, 1958, 1965; Kuwabara, 1968, 1970, 1975; Kuwabara *et al.*, 1969; Kuwabara and Imaizumi, 1974; Gorthy, 1968; Gorthy *et al.*, 1971; Hogan *et al.*, 1971; Leeson, 1971; Dickson and Crock, 1972, 1974; Rafferty, 1972; Rafferty and Esson, 1974; Rafferty *et al.*, 1974; Harding *et al.*, 1973; Wakely, 1974; Rafferty and Goosens, 1975; Hamai *et al.*, 1974; Hamai and Kuwabara, 1975; Farnsworth *et al.*, 1974, 1976; Sukuragawa, 1975; Hollenberg *et al.*, 1976; Nelson and Rafferty, 1976; Benedetti *et al.*, 1976). This list is by no means complete. The citations listed are only those consulted in the preparation of this review.

A. Capsule

Each of the lenses is entirely surrounded by a tough, collagenous capsule which is thicker on the anterior side than on the posterior (Fig. 1). The degree of this thickness disparity differs from species to species. This simple difference in capsule thickness has required that the majority of electrophysiologic measurements made with fine glass-capillary microelectrodes be made through the posterior surface. The thicker anterior capsule frequently causes the microelectrode to break when anterior penetration is

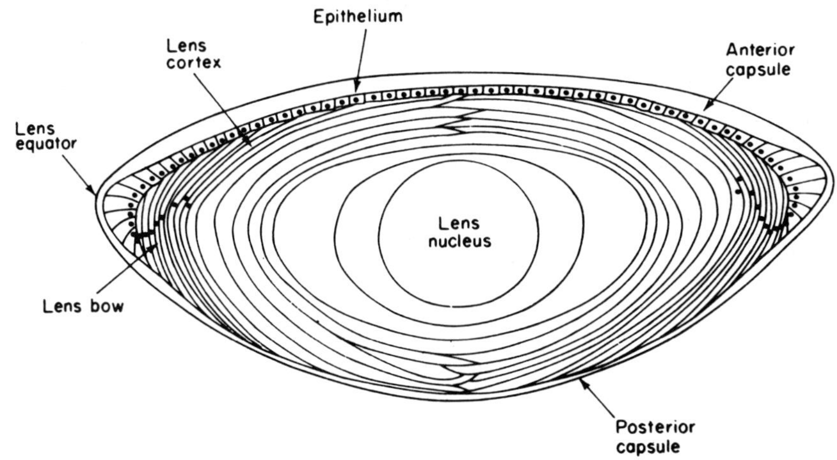

FIG. 1. Schematic drawing of the structural organization of the dog lens. (From Pirie and Van Heyningen, 1956.)

attempted. The existence of the capsule has also prevented routine penetration of anterior epithelial cells or surface posterior fibers, each of which lies directly under the capsule. When the electrode encounters the capsule, capsular dimpling occurs. If the electrode is continually advanced, it finally penetrates the capsule but comes to rest usually 70–100 μm below the lens surface, a depth greater than that of either surface fibers or epithelial cells. Penetration of the surface cells upon electrode withdrawal would seem unsatisfactory because of the high probability that these cells would have been damaged by the initial penetration. It is known that the lens capsule can be removed enzymatically by the use of collagenase (Pirie, 1951; Becker and Cotlier, 1965; Kern, 1965; Duncan, 1969b; Paterson, 1970c). The combination of this decapsulation technique with microelectrode methods offers a promising approach to the eventual understanding of the electrophysiologic role of surface lens cells.

B. Epithelium

The lens is morphologically asymmetric in that it possesses a cuboidal-type epithelium on its anterior side while lacking such an epithelium on its posterior side. These epithelial cells apparently have similar ultrastructure to epithelial cells from other tissues. For example, there are moderate numbers of mitochondria (Wanko and Gavin, 1958; Gorthy, 1968) that occur more frequently in the equatorial region than in the intermediate or central zones (Wanko and Gavin, 1958). Location of mitochondria is of interest since it is well known that enzymes responsible for aerobic metabolism are generally associated with the mitochondria in most tissues. Therefore, the aerobic metabolism known to occur sparingly in the lens (Kinsey and Frohman, 1951) must occur primarily in the epithelium and may be asymmetrically distributed. The epithelial cells are nucleated and have abundant ribosomes and endoplasmic reticulum suggesting they are capable of enzyme and structural protein synthesis. A well-developed Golgi apparatus usually lying between the epithelial cells nucleus and the fibers below is a frequent finding, suggesting a secretory function for the epithelial cells. These cells are thought to be responsible for the secretion of the capsule (Young and Ocumpaugh, 1966). Microtubules are also found in the epithelial cells as well as in some differentiating cells at the equatorial region (Kuwabara, 1968). The function of these microtubules in the lens is not at all clear.

C. Lens Fibers

The fiber cells that make up the majority of the lens are formed from nucleated epithelial cells in the equatorial region of the lens. As these epithelial cells differentiate into fibers, they lengthen considerably and eventually lose their nuclei and other intracellular organelles. This loss of organelles has been somewhat overemphasized in the past. Many fibers can be found that have essentially the same organelles, although less abundant, as those in epithelial cells. In fact, nuclei can routinely be found in fibers 30–50 layers deep in most lenses. The loss of organelles, especially the nucleus, is a relatively slow process. The differentiating cells lengthen in such a way that the part analogous to the outside (closest to capsule) surface of the undifferentiated epithelial cells ends up on the posterior side of the lens. The part analogous to the inner surface ends up on the anterior side. These fibers continue to lengthen until they meet cells elongating from other regions of the lens. Where the fiber cells meet, both anteriorly and posteriorly, they form a complex radiating pattern of lines referred to as the anterior and posterior sutures. There is substantial fiber interdigitation in these suture areas. Fowlks (1965) has suggested that substances from the bathing solution might more easily penetrate the lens extracellular space in the sutural region than elsewhere, a point that apparently has not been pursued. Each new fiber cell is laid down to the outside of the next oldest fiber cell. This action causes the inner fibers to be pushed in toward the center of the lens, presumably compressing them more and more to form the lens nucleus. The lens never loses cells to the outside world and so contains every cell it has produced. This growth pattern leads one to question if cells deep in the lens are alive and important to ion metabolism and electrophysiology. If not, at what level do they cease to be important? These questions are at present unanswered, but represent important areas for future research.

D. Cell-to-Cell Junctions and Interdigitations

1. General Review

Perhaps the most important specialized morphologic structures in the lens from the electrophysiologic standpoint are its specialized cell-to-cell junctions. An excellent review of lens cell junctions has recently been published (Benedetti *et al.*, 1976). Such junctions are ubiquitous in most epithelial, endothelial, and nervous tissues (Fahrquar and Palade, 1963). Consequently, it would seem useful to briefly review such junctions in general. Reviews in more detail exist elsewhere (Gilula, 1974; Pappas and Purpura, 1972; McNutt and Weinstein, 1970, 1973; Weinstein and McNutt, 1972; Bennett, 1973a; Benedetti *et al.*, 1976).

A cell junction is a region of short-range contact between two cells where the ultrastructure of the adjacent plasma membrane differs from that of the general plasma membrane. There may also be differences in the intercellular matrix associated with these specialized membrane areas. Though many such cell junctions have been described, only three appear to have a potential electrophysiologic function. These are the zonula occludens, macula occludens, and gap junction. For each junction, casual examination under the electron microscope makes it appear that the adjacent cell membranes fuse in the area of the junction. In spite of their similar appearance, these junctions are thought to have very different functions. Thus, it is critical to be able to distinguish between them morphologically.

The zonula occludens is a continuous, beltlike region of intimate contact between the plasma membranes of adjacent cells. This junction completely rings the cells in question. The macula occludens has essentially the same structure except that it occurs in intermittent, short segments rather than in a continuous belt. In thin section, the gap junction can look either identical to both macula and zonula occludens or very different from each depending on the protocol used for the specimen preparation. In many tissues, the zonula occludens forms a seal between adjacent cells, preventing extracellular diffusion of certain types of molecules from one side of the cell to the other (Fahrquar and Palade, 1963; Revel and Karnovsky, 1967; Karnovsky, 1968; Goodenough and Revel, 1970). Such junctions are common between epithelial cells or other cells separating the lumen of cavity organs from structures lying in the wall or on the opposite side of the wall. The maculae occludentes may regulate the size of molecules that are allowed to diffuse around the cells via the extracellular space. By regulation of the spacing of the maculae occludentes, "pores" may be created around cells which allow diffusion of the molecules which do not penetrate the cell membranes (Karnovsky, 1968). The gap junction is thought to provide adhesion and direct communication between cells (Lowenstein, 1966; Bennett, 1973a). Such communication allows the direct movement of low-molecular-weight substances directly from one cell to another without entering the extracellular space (Bennett, 1973b). Gap junctions, because they allow ready movement of ions, appear to be responsible for electrical coupling between cells (Lowenstein, 1966). The maximum size of molecule that can pass is at least 10 Å, but may be larger (McNutt and Weinstein, 1973).

There are three basic methods for morphologically distinguishing occluded vs. gap junctions (Gilula, 1974). In the first of these, after fixation with glutaraldehyde and osmium, the tissue is stained en bloc with uranyl acetate (Revel and Karnovsky, 1967). In thin cross section, the gap junction then appears to have seven layers. The seven layers result from the parallel apposition of two unit membranes, each 60–75 Å thick, separated by a 20–40 Å space or gap. The total junctional thickness is 140–190 Å, more than 2× the thickness of a single unit membrane. Occluded junctions preserved in this manner would lack the gap and would thus appear pentalayered. They would have a total thickness of 110–150 Å, less than 2× the thickness of a single unit membrane. In the second procedure, an electron-opaque tracer such as colloidal lanthanum hydroxide (Revel and Karnovsky, 1967) is placed in the solution around the tissue either before and/or during fixation or after cutting into blocks. Such material readily penetrates the "gap" showing it to be in communication with the extracellular space (Fig. 2b,c). Other substances such as pyroantimonate or ruthenium red may also be used. Since the macula or zonula occludens lacks a gap in communication with the extracellular space, no tracer can be found in such junctions. Finally, freeze-fracture techniques (Moore, 1966; McNutt and Weinstein, 1973; Gilula, 1974) can distinguish gap and occluded junctions. When an occluded junction is cleaved, one surface shows long ridges, whereas the complementary surface shows long corresponding grooves (McNutt and Weinstein, 1973). When a gap junction is cleaved, one fracture surface shows myriad 60–80 Å unit particles, whereas the other shows 30–40 Å pits (Fig. 2a). The particles are often tightly packed and frequently in a hexagonal array with 90–100 Å center-to-center spacing. A 20–25 Å depression is often seen on top of the particles. It has been speculated that each of these depressions is the site of a hydrophilic channel joining the two cells (McNutt and Weinstein, 1970). In some cases, negative staining has also been found useful in

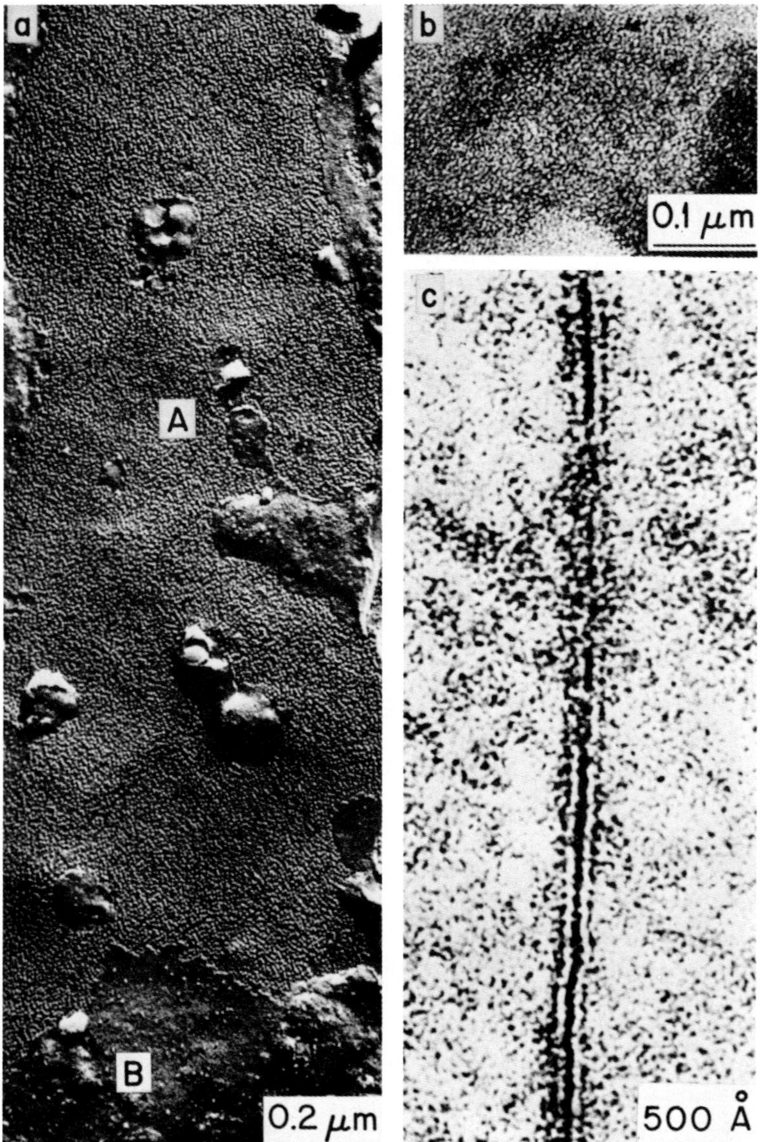

FIG. 2. Gap junctions between lens cortical fibers. (a) Freeze-fracture of extensive junctional domain. Fracture face A shows junctional particles whereas fracture face B shows corresponding pits. (b) and (c) Tangential and cross sections of cortical junctions impregnated by lanthanum hydroxide. (From Benedetti *et al.*, 1976.)

identifying gap junctions (Gilula, 1974). Although this discussion may have made it seem relatively straightforward to morphologically distinguish between gap and occluded junctions, one should realize that these junctional morphologic features are not invariant (Friend and Gilula, 1972). Therefore, it is frequently difficult to unequivocally identify the type of junction with which one is dealing.

2. Lens Epithelial Junctions

Lens epithelial cells have been reported to show occluded-type junctions between their lateral walls (Wanko and Gavin, 1958; Cohen, 1965; Gorthy, 1968; Gorthy *et al.*, 1971; Kuwabara *et al.*, 1969; Hogan *et al.*, 1971; Rafferty *et al.*, 1974; Rafferty and Esson, 1974; Kuwabara, 1975). As previously suggested, such junctions are usually found in epithelia to limit the movement of substances across the epithelium by way of the extracellular channels. If these junctions were to completely ring each epithelial cell (zonula occludens), they could make the extracellular space between these cells a very high resistance pathway for the movements of ions and other substances. It would appear, however, that the occluded junctions between epithelial cells may not completely ring the cells. Cohen (1965) found that in human lenses, they were too variable in their position to form complete "seals" around the cells and suggested they occurred in patches. He further suggested that their role might be to provide cell-to-cell adhesion. Gorthy *et al.* (1971) found that horseradish peroxidase penetrated the extracellular space between the epithelial cells to reach the anterior cortex in rat lenses. The anterior penetration was less than either that found at the equator or the posterior surface of the lens. The fact that it penetrated at all probably means that the seal around epithelial cells is not complete. In our experience, it is relatively easy, in rat and frog lenses, to find lateral epithelial cell borders where junctions are completely absent. This again suggests that there cannot be a good seal around epithelial cells (Rae and Stacey, unpublished observations). Benedetti *et al.* (1976) did not find occluded junctions between epithelial cells. Unfortunately, there are few studies in the lens that attempt to quantitate the relative distribution of such lateral occluded junctions from anterior pole to equator, or to assess the extent to which the occluded junctions "ring" the epithelial cells. Also, there are few studies that attempt to look at the ability of these occluded junctions to limit the movement of extracellular tracers. Further studies would be welcomed and necessary if one is to understand any "barrier" function which the lens epithelium might have. Lateral junctions between adjacent epithelial cells may be gap junctions. Some investigators even call them gaps (Rafferty *et al.*, 1974; Rafferty and Esson, 1974; Kuwabara,

1975). Benedetti *et al.* (1976) found real gap junctions between epithelial cells, using freeze-fracture techniques. The frequency of occurrence of these junctions is not yet clear. There are few unequivocal physiologic or morphologic studies that demonstrate "coupling" between significant numbers of adjacent epithelial cells.

In addition to these occluded junctions on the lateral walls, similar junctions also occur between the base of the epithelial cell and the fiber layer immediately below (Wanko and Gavin, 1958, 1959; Cohen, 1965; Gorthy, 1968; Gorthy *et al.*, 1971; Hogan *et al.*, 1971; Kuwabara *et al.*, 1969). Hogan *et al.* (1971) point out that these junctions occur very frequently, whereas other investigators do not describe the frequency of their occurrence. In our experience in rat lenses, they occur infrequently. Whether these are occluded junctions or gap junctions is still in question. All of the early literature referred to such junctions as occluded junctions. These early studies were published, however, before the morphologists realized that there was often a clear distinction between the two types of junctions. These epithelial–fiber junctions, if they are gap junctions, become of potentially great significance physiologically. As reported earlier, it is well known that gap junctions represent a structure that provides cytoplasmic continuity between adjacent cells. In such junctions, varying numbers of "cytoplasmic tubules" having a lumen diameter of approximately 10 Å units are thought to extend directly from the cytoplasm of one cell to that of an adjacent cell. These cytoplasmic bridges allow ions and small-molecular-weight substances (<1000) to pass directly from one cell to another without entering the extracellular space. That is, they provide low-resistance communication between cells. Such gap junctions in the lens would provide channels for the movement of substances directly from the epithelial cell into the fibers below and vice versa. There are two possible pathways by which materials might get from the bathing solution into the fiber cells of the lens. One is to pass between the epithelial cells to enter the extracellular space of the lens and from there cross the fiber membranes into the cytoplasm of the fiber itself. The second would be to enter the epithelial cell from the bathing solution, traverse the cytoplasm of the epithelial cell, and enter the fiber through the gap junction between the epithelium and fiber cell. It is of great importance to know the relative contribution of these two pathways. At the present time, however, no direct studies exist to deal with this important problem.

The extracellular pathway around the epithelial cell is apparently a long and tortuous one. Most electron micrographs of lens epithelial cells show considerable lateral interdigitation between adjacent cells. Such interdigitation is found to be more extensive as one approaches the equatorial region where the cells become more columnar (Wanko and Gavin, 1968).

Scanning electron micrographs (Farnsworth *et al.*, 1974) clearly show myriad interdigitating, fingerlike projections extending between adjacent epithelial cells. In experiments dealing with penetration of substances into the lens, such anatomical structures might make it appear that the epithelial cells were providing a significant diffusion barrier. Alternatively, the observed reduced penetration might result completely or in part from the increased diffusion path length. One should view this statement cautiously, however, since a real epithelial barrier provided by occluded junctions cannot be entirely ruled out at this time.

3. Lens Fiber Junctions

The lens fiber cells that make up the majority of the lens are very tightly packed on all sides. They assume a hexagonal array, thus maximizing the packing and minimizing the extracellular space. Morphologic studies utilizing the scanning electron microscope have shown that a considerable number of interdigitations occur between the lens fibers (Kuwabara, 1970, 1975; Yamasaki, 1970; Hansson, 1970; Dickson and Crock, 1972, 1974; Matsuto, 1973; Farnsworth *et al.*, 1974; Nelson and Rafferty, 1976; Hollenberg *et al.*, 1976; Harding *et al.*, 1976) (Fig. 3). Farnsworth *et al.* (1974) in rat lenses, showed three distinct variations in fiber structure depending upon the area of the lens being studied. In the anterior cortex, the fibers were interconnected by numerous ball and socket junctions on each plane or surface. In the lateral part of the lens, numerous fiber interdigitations occurred, more frequent in rows along the edge of the fiber. The posterior fibers appeared more rectangular in cross section and had fewer interlocking projections. The significance of this variable fiber-to-fiber morphology is unknown and will not be discussed. This very complex geometry is significant to the electrophysiologist from two standpoints. First, it is in principle possible to calculate the volume of the extracellular space from electron micrographs assuming some regular consistent geometry. However, any calculations that do not take this extensive interdigitation into account would considerably underestimate the extracellular space. Also, measurements made from electron micrographs in one area would not be representative of other areas of the lens since the extent of interdigitation varies substantially from one region to another. Second, it would be beneficial to know the total surface area of all fibers composing the lens. Intact cell membranes will have a resistance and capacitance associated with them. In studies in which current is passed through the lens and the resulting change in the lens potential is examined, the time course of this voltage change should be related to the total capacitance and thus the total surface area of the compartment being charged. It is therefore of some interest to know a precise figure for the total surface area. The

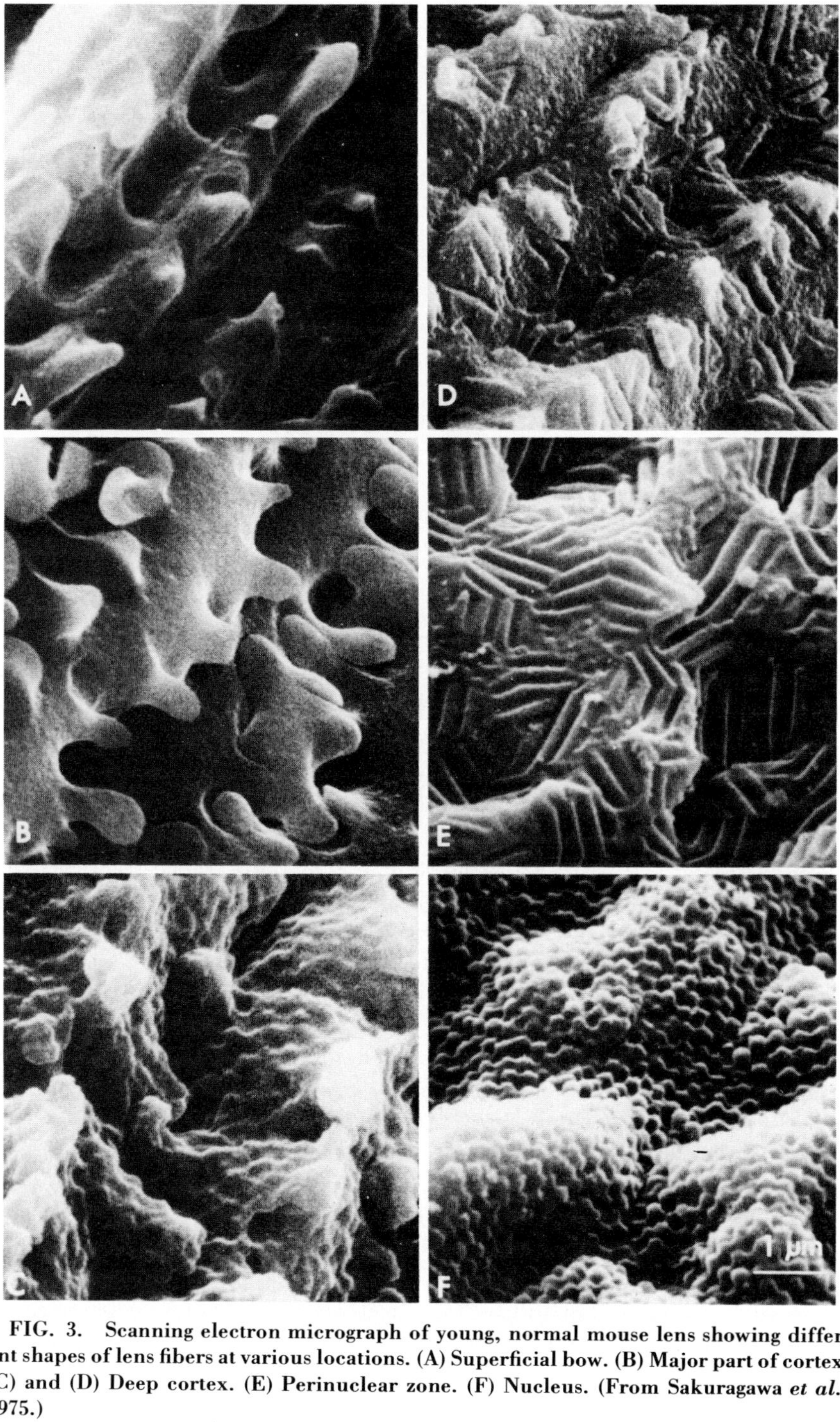

FIG. 3. Scanning electron micrograph of young, normal mouse lens showing different shapes of lens fibers at various locations. (A) Superficial bow. (B) Major part of cortex. (C) and (D) Deep cortex. (E) Perinuclear zone. (F) Nucleus. (From Sakuragawa *et al.*, 1975.)

complexities due to fiber interdigitation would seem to make extracellular space and total surface area difficult or impossible to determine from electron micrographs.

Electron micrographs have shown the fibers to be so tightly packed that only 80–100 Å channels exist between adjacent cells. Since lens microelectrode studies have been done with electrodes having tip diameters of around 1 μm (10,000 Å), it is clear that the electrode tip could not routinely end up in the extracellular space. This size disparity could explain why most investigators have been unable to detect a change in potential as the electrode left a fiber and entered the extracellular space in the lens. It would seem that when a microelectrode is placed in the lens, the tip comes to rest in a fiber and that it is difficult to make it do otherwise.

The lens fibers also show numerous cell-to-cell junctions. Both gap and occluded junctions have presumably been identified, but the relative percentage of occurrence and distribution of each is unknown. Isolated membrane fractions from the lens (Lasser and Balazs, 1972; Bloemendal *et al.*, 1972; Dunia *et al.*, 1974; Benedetti *et al.*, 1974, 1976) contain junctional complexes that appear pentalaminar, a fact that suggests they are occluded junctions. However, freeze-fracture studies show that the junctions lack the ridges and furrows associated with occluded junctions, and thus cannot be tight junctions (Bloemendal *et al.*, 1972). Also, the hexagonal lattice, with 90 Å center-to-center repeating subunits characteristic of gap junctions from other tissues seen in freeze-fracture, was not consistently observed. Bloemendal *et al.* (1972) suggested that the junctions were neither occluded junctions nor the usual variety of gap junction. Nonaka *et al.* (1976) reported that in their freeze-fracture lens preparations the majority of the fiber junctions were gap junctions. Phillipson *et al.* (1975) and Rae and Stacey (1976) reported that gap junctions (seven layered) could be found in human, bovine, and rat lenses en bloc stained with uranyl acetate. The relative occurrence of these junctions with regard to the different surfaces of the hexagonal fibers has not been investigated. The occurrence of occluded junctions between fibers that might limit the movement of materials in the extracellular space has not yet been described (Benedetti *et al.*, 1976). Either a gap junction or an occluded junction might provide mechanical strength for a lens changing its shape during accommodation. Kuwabara *et al.* (1969) have shown that in swollen, galactose cataractous lenses, the junctional areas between cells continue to adhere tightly. A function other than mechanical rigidity is suggested by the fact that fish lenses which do not appreciably change their shape on accommodation still contain large numbers of these fiber-to-fiber junctions (Wakely, 1974). However, even in fish lenses a structural role for the junction cannot be excluded. Thus, it seems that lens fibers do contain many cell-to-cell junctions that are gap junctions by some criteria, but not by others. Rae (1974a) and Rae and Stacey (1976) have shown that Procion dyes can pass between fibers that possess such junctions. One cannot rule out the possibility that these junctions in the lens might be specialized to allow even freer cell-to-cell movement of diffusing substances than found elsewhere. Such a specialization would seem beneficial to the lens, but its existence is highly speculative at this moment (Benedetti *et al.*, 1976).

E. Deep Fiber Membranes

There is some question that the membranes of fibers deep in the lens are functionally intact. Morphologic data on membrane intactness are equivocal. Electron microscopically, the cells in the superficial cortex seem to have obvious fiber membranes. However, as one proceeds deeper into the lens, it becomes more and more difficult to appreciate continuous fiber membranes under the electron microscope. It is not clear whether this lack of membranes is real or if it represents the inability of fixatives to adequately penetrate the densely packed lens material. Cohen (1965) and Hogan *et al.* (1971) have commented on these fixation problems. Hogan *et al.* (1971) have shown micrographs of the fibers of the human lens nucleus in which the membranes appear to be continuous. The results of most investigators (Kuwabara, 1968, 1975; Kuwabara *et al.*, 1969; Wakely, 1974; Phillipson *et al.*, 1975; Rae and Stacey, unpublished observations) have shown that it is relatively easy to identify isolated segments of morphologically normal membrane there. However, normal membrane is frequently not continuous around the entire cell in the nuclear region. Pauly and Schwan (1964) have electrical impedance data suggesting deep fiber membranes might not be intact. The existence of membranes throughout the lens, as well as specifically in the nucleus, has recently been verified by optical methods (Bettelheim and Vinciguerra, 1971; Vinciguerra and Bettelheim, 1971; Bettelheim *et al.*, 1973; Bettelheim and Wang, 1974; Bettelheim, 1975). Such optical studies cannot prove, however, that the membranes identified have high electrical resistance or possess ion pumps. Consequently, the question concerning the intactness of deep cortical and nuclear fiber membranes remains.

III. EXTRACELLULAR SPACE

A. Electron Microscope Studies

Adequate analysis of radioactive tracer methods requires that the volume and distribution of the extracellular space be known. As stated earlier, electron micrographs of the lens have shown the extracellular space to be very abbreviated (Cohen, 1965). On the basis of electron micrographic data, some investigators (Kinsey and Reddy, 1965) have even suggested that perhaps no extracellular space exists. Cohen (1965), by comparing the extracellular dimensions in the crystalline lens to that found in brain tissue, suggested that the extracellular space might be around 5–15%. Paterson (1970a), from electron micrographic photographs, calculated the extracellular space in the cortex to be approximately 0.8%, but did not take into account cell-to-cell interdigitations. He, however, noted the importance of repeating the measurements so as to include an interdigitation factor. Because of the fixation difficulties that all lens morphologists have experienced and because of the high probability that the lens tissue is altered by the histologic preparation involved, it is unlikely that one would gain an accurate estimate of the extracellular space from electron micrographs. Such an estimate is further complicated by the fact that the complexity of cell-to-cell interdigitations differs markedly from one area of the lens to another.

B. Isotope Studies

Extracellular space measurements are probably best made using radioactive, extracellular space markers. Ideally, an extracellular marker should be one that does not penetrate into the cytoplasm of the lens cells, does not bind to any extracellular structure, is of such dimensions that it penetrates even the smallest extracellular niches, diffuses rapidly so that extended time measurements are avoided (Paterson, 1970a), and can be located using morphologic techniques. Certainly, no such ideal substances exist at the present time. Therefore, the measurement of the extracellular space in the lens is fraught with uncertainties. Early measurements of lens extracellular space were made by Huggert (1959a,b) using ^{24}Na. He found values as high as 35% depending on the conditions used for the determination, but reported the normal *in vitro* ^{24}Na space to be 10–12% in the cortex and 4–7% in the nucleus. It is now well known that Na^+ penetrates most cells to some degree and as a result is not an appropriate extracellular tracer. Huggert (1961) and later Duncan (1970b) made autoradiographs of lenses incubated in ^{22}Na and showed that the Na^+ did not penetrate into the nucleus even after 6–48 hours (Fig. 4). One might surmise that the Na^+ appreciably penetrated into the lens fibers since the autoradiographs show it to be distributed ubiquitously throughout the cortex and not just in the extracellular space. The resolution of ^{22}Na autoradiography is not very good, however, so one should be cautious about that interpretation. Duncan (1970b) and Paterson (1970c) have shown by Na^+ efflux measurements, that Na^+ penetrates into the lens fibers. Such efflux curves show a rapid,

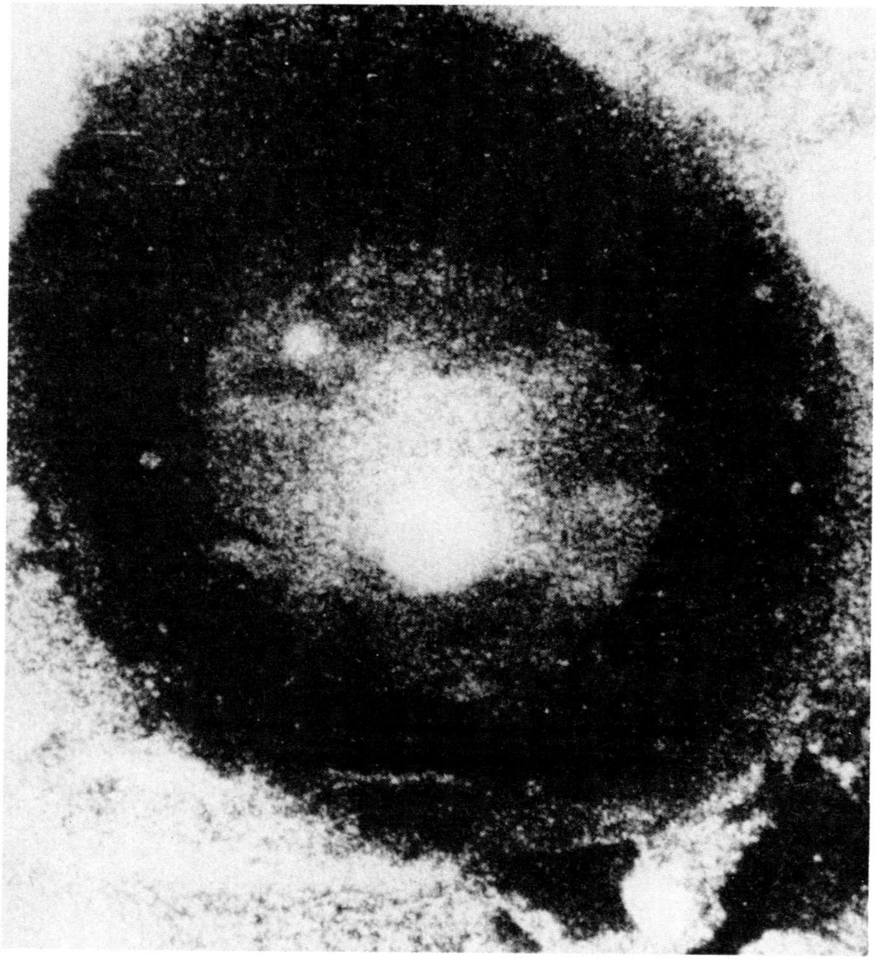

FIG. 4. Autoradiograph of ^{22}Na penetration into toad lens after 6 hours of incubation. (From Duncan, 1970b.)

presumably extracellular component, followed by two, longer, time-constant components probably due to efflux from the intracellular compartment.

Thoft and Kinoshita (1965a) showed that the extracellular space of the rat lens depended very much on the measuring molecule used. Their ^{24}Na space was about 13% of the lens water, the ^{14}C-inulin space was only about 5%, with ^{14}C-mannitol and ^{14}C-sucrose falling in between at about 7–8%. The differences could have been due to the large inulin molecule's inability to penetrate some of the smaller extracellular channels. Alternatively, mannitol, sucrose, and Na^+ might be able to penetrate the cell membranes to some degree and thus gain access to the intracellular space as well. Paterson (1968) even showed the ^{14}C-inulin space to be different for different commercially available inulins. Later, he (1970a) reviewed the majority of the measurements made on extracellular space and thoroughly discussed the problems involved in measuring extracellular space in the lens. He found the ^{24}Na or ^{22}Na space to be about 10% of lens water, whereas the ^{14}C-inulin space was only about 5%. Importantly, he showed that in rabbit lenses, 21% of the extracellular space was located in the capsule, 66% in the cortex, and approximately 13% in the nucleus. His ^{14}C-inulin penetrated into the nucleus of the lens within 6 hours, a surprising finding since both Huggert (1961) and Duncan (1970b) had found autoradiographically that ^{22}Na did not. Whether or not the ^{14}C-inulin reaches the nucleus via the extracellular space is a matter yet to be determined since no autoradiographic measurements of the location of inulin in the lens have been performed. This discrepancy deserves more attention since it raises the important question: is there any extracellular space in the nucleus of the lens? Certainly if one surrounds the lens with a bathing medium containing radiolabeled ions, carbohydrates, or amino acids, waits a suitable period, subsequently peels the lens apart, and assays for radioactivity in the nucleus, such radioactivity is invariably found. Becker and Cotlier (1962) found ^{86}Rb reached the nucleus after 30 minutes of incubation, in comparison to Paterson's 6 hours for ^{14}C-inulin. Such studies do not show how the labeled substance got there. Since the lens fibers are coupled by gap junctions, both substances might have gotten to the nucleus by entering the lens fibers and diffusing from one cell to another through the junctions. No studies, at this time, unequivocally show movement of tracers in the extracellular space of the nucleus. Clarification of this question would be welcome.

Duncan (1970b), using ^{14}C-sucrose, has shown the extracellular space to be only about 2% of the lens water. In these measurements, he allowed the lens to become loaded with ^{14}C-sucrose and then looked at the sucrose efflux. There was an early rapid component followed by a much longer component. Duncan assumed that the rapid component represented the efflux from the true extracellular space, whereas the slow component represented efflux of sucrose that had penetrated the lens fibers. Again, no morphologic evidence of the location of the labeled material was presented. Since Paterson (1970a) and, more recently, Yorio and Bentley (1976) have shown that an appreciable fraction of the extracellular space exists in the capsule of the lens, an appreciable portion of Duncan's 2% extracellular space must also be located in the capsule. This means that the extracellular space of the cortex and the nucleus would be even less than that reported by Duncan and would begin approaching the results of electron microscopic measurements. Most recently, Yorio and Bentley (1976) have measured the extracellular space in the amphibian (*Bufo marinus*) lens. By comparing ^{14}C-inulin space determinations before and after decapsulation of the lens with collagenase, they concluded that 80% of the extracellular space in the toad lens existed in the capsule. They reported a ^{14}C-inulin space of 7% in the intact lens, whereas it was only 1.2% in the decapsulated lens. This value of 1.2% again would correlate reasonably well with the electron microscopic measurements especially when one considers that to date no attempts have been made to account for the additional extracellular space that would occur around the numerous interdigitations in the lens fibers and epithelial cells. Yorio and Bentley (1976) also placed the toad lens in an Ussing chamber (Ussing and Zerahn, 1951) and presented the inulin or mannitol exclusively to either the anterior or posterior side of the lens. They found that during the 3-hour incubation time, more of the labeled substance entered the lens from the posterior surface than from the anterior surface. The authors suggested that the epithelial cells must present a "barrier" to the penetration of these large molecules. Thoft and Kinoshita (1965a) also found the sucrose space determined *in vitro* to be more than double that determined *in vivo*, suggesting possible damage to the lens in the *in vitro* condition. The *in vivo* space was measured using ^{14}C-sucrose which was injected intravenously. Shortly thereafter, labeled sucrose was found to be much more concentrated in the aqueous humor than in the vitreous body. Consequently, the extracellular space was determined more by movement of the label across the anterior surface than the posterior. Yorio and Bentley's data would suggest that the extracellular space so determined would be less than for total lens immersion, unless, of course, one waited long enough to ensure that complete equilibration of the isotope in the lens had occurred. Thoft and Kinoshita themselves suggested an error in the *in vivo* space might have occurred, not by lack of equilibrium, but by assuming in their calculations that the lens was surrounded by a fluid having a ^{14}C-inulin concentration equal to that in the aqueous humor. This assumption would certainly underestimate the extracellular space, but it is impossible to quantitate the degree of underestimation. The *in vivo–in vitro* discrepancy is important for, if real, it means that a significant alteration occurs in the lens when it is incubated *in vitro*. This fact would have to be considered when one interprets results from *in vitro* experiments.

C. Summary

In summary then, we are left with great uncertainties about the size and distribution of the extracellular space. One can take the size of the sodium space at about 10–13% to represent the upper limit, realizing that it is surely too large. The ^{14}C-inulin space, because of the large size of the molecule limiting its access to some extracellular channels, might be too small. We cannot, however, be certain that inulin does not penetrate into the lens cells. Paterson (1970a) and Duncan (1974) have pointed out that inulin is known to penetrate the membranes of other cells (Ling and Kromash, 1967; Brown *et al.*, 1969). From this standpoint, the inulin space might be too large. Paterson (1970a) reports that 20% of the extracellular space is in the capsule in rabbit lenses, whereas Yorio and Bentley (1976) say that 80% of the space is in the capsule in amphibians. All electron microscopic measurements, as well as the flux data of Duncan (1970b) and Yorio and Bentley (1976), suggest an extracellular space in the cortex–nucleus combination of approximately 0.8–1.5%. Paterson's data, after correction for the amount in the capsule, would amount to about 4% for ^{14}C-inulin. Therefore, we know that the extracellular space in the lens is small, but certainly exists. The size of the extracellular space of the nucleus must remain in question until someone unequivocally shows the labeled marker in the nuclear extracellular space. It is clear, however, that the size of the nuclear extracellular space is less than that of the cortex. The quantitative values of any of the spaces calculated to date should be considered cautiously since, whether the extracellular space is measured by chemical analysis or by isotope tracer analysis, the assumption must be made that the extracellular concentration of the substance in question is identical to that in the bathing solution surrounding the lens. Since the permeabilities of the clefts between epithelial cells and between posterior lens fibers have not yet been determined, nor has the relative ability of substances to diffuse in the tortuous channels between lens fibers been determined except near the lens surface (Paterson and Maurice, 1971), such an assumption is open to question. The fact that some extracellular space can be demonstrated is important since it means that a pathway for solute movement and current flow exists between at least some lens fibers. The physiologic significance of this pathway is yet to be determined.

IV. ELECTROPHYSIOLOGY

A. Fiber Potential vs. Outside Layer Potential

Brindley (1956) was the first to measure the potential difference (PD) of the lens with glass microelectrodes. He made measurements from both the rabbit lens and the frog lens and determined that the inside of each lens was maintained at a potential approximately 70 mV more negative than the bathing solution. He found that all areas of the lens were isopotential or at least had very similar PDs. A controversy soon arose: Was the lens acting simply like a giant cell with the PD being developed only across an outside cell layer of the lens, or did each of the individual fibers inside the lens also maintain a PD across its membranes? In light of the asymmetric anatomy of the lens, it is not clear why this controversy arose since it would at the outset seem rather farfetched to consider the PD to exist across only an outside layer of cells. However, even today, the controversy is not completely resolved. Sperelakis and Potts (1959) found that when they penetrated the decapsulated bovine lens with glass microelectrodes, 20–25 mV depolarizing PD jumps were seen which they reported to be PDs across individual fiber membranes. Andree (1958a), using microdissection and microelectrode techniques, reported the PD existed only across the outside layer. Duncan (1969a, 1969b) supported Andree's viewpoint. He arrived at this conclusion from three different pieces of information.

1. When he penetrated the amphibian lens with glass microelectrodes, he did not see jumps in the PD which would correlate with the existence of fiber membranes.

2. He step-depolarized the lens by quickly superfusing the surface with solutions having elevated potassium chloride concentrations while he simultaneously measured the glass microelectrode PD at depths of up to 500 μm below the surface. His analysis of this data was based on a previous approach by Dainty and House (1966) who used frog skin. Basically the approach assumes that the half-time for the PD change, following a change in the bath KCl concentration, should be related to how far the K^+ has to diffuse to reach the site of the PD producing membranes, i.e., the diffusion path length. Quantitatively, $t_{\frac{1}{2}} = 0.38\delta^2/D_k$ (Dainty and House, 1966), where $t_{\frac{1}{2}}$ is the time for the K^+ concentration adjacent to the membrane to reach a value $K^+_{\frac{1}{2}} = 0.5\ [K^+_1 + K^+_2]$ (K^+_1 is the original concentration in the bulk fluid and K^+_2 is the final concentration), δ is the diffusion path length for K^+ in centimeters, and D_k is the diffusion coefficient of K^+ in the unstirred layer in square centimeters per second (Duncan, 1969b). Duncan found the PD to change far too rapidly for the K^+ to have diffused all the way to the location of the microelectrode tip. In fact, it appeared that the K^+ had only to diffuse across an unstirred layer, the capsule, and one or two layers of lens cells to account for the time course of the PD change. This result, along with Andree's result that the lens PD was markedly reduced by disrupting only surface layers, made it clear that the PD could not be due simply to independent PDs across individual fiber membranes. Alteration of surface cells or their environment would not be expected to alter the PD of deeper fibers if each fiber maintained an independent PD.

3. Finally, he placed two glass microelectrodes in the lens, one for passing current and the other for measuring the PD induced by the flow of the current across the lens "membrane" resistance. He repeated these measurements for different separations of the current and voltage microelectrodes and found that the induced PD was independent of the electrode depth. He reasoned that if each fiber membrane had been adding to the resistance across which the induced voltage was generated, this spatial independence of the induced PD would not have been found. The circuit diagram which he envisioned to arrive at this line of reasoning is not clear. His exciting finding was that he was able to record an induced PD for all separations of the electrodes, even when it was certain that the electrodes could not be in the same cell. In retrospect, this was the first direct evidence for cell-to-cell coupling in the lens.

Duncan, however, concluded that the PD existed only across an outside layer of cells and that the interior membranes were degenerate perhaps due to lysolecithin within the lens which might have caused membrane disruption.

Rae and Blankenship (1973) performed three different experimental procedures which they reported did not support the concept of internal fiber degeneracy. In the first of these, they used double-barrel microelectrodes, i.e., two microelectrodes cemented together with the tips only a fraction of a micron apart, for measuring the input resistance of individual lens fibers. Both tips were placed within a single cell, current was passed through one, and voltage was measured by the other. Rae and Blankenship reported input resistance measurements of 1–6 MΩ within a single lens at different locations. In addition, assuming a membrane capacitance of 1 $\mu F/cm^2$, they found that the charging occurred far too fast for an entire layer around the outside of the lens to have been charged. Therefore, they hypothesized that they were charging only individual fibers. The conclusion was very much in error since it did not consider the cable properties of the lens. In retrospect, the measurements did demonstrate the existence of a high convergence resistance in the region of the tip (Eisenberg and Rae, 1976) suggesting the existence of a resistive substance (presumably membrane) other than cytoplasm alone. In the second set of experiments, they impaled the lens with a microelectrode filled with Procion yellow or Procion scarlet dye materials. They subsequently iontophoretically placed the negatively charged dye within the fiber and reported the diffusion pattern which they saw was consistent with the dye remaining within a single fiber (Fig. 5). They concluded that since the dye was contained within the fibers, the membranes must be intact. In their third type of experiment, a glass microelectrode was placed into the lens and then slowly placed deeper and deeper in 1 μm steps using a high precision hydraulic micromanipulator

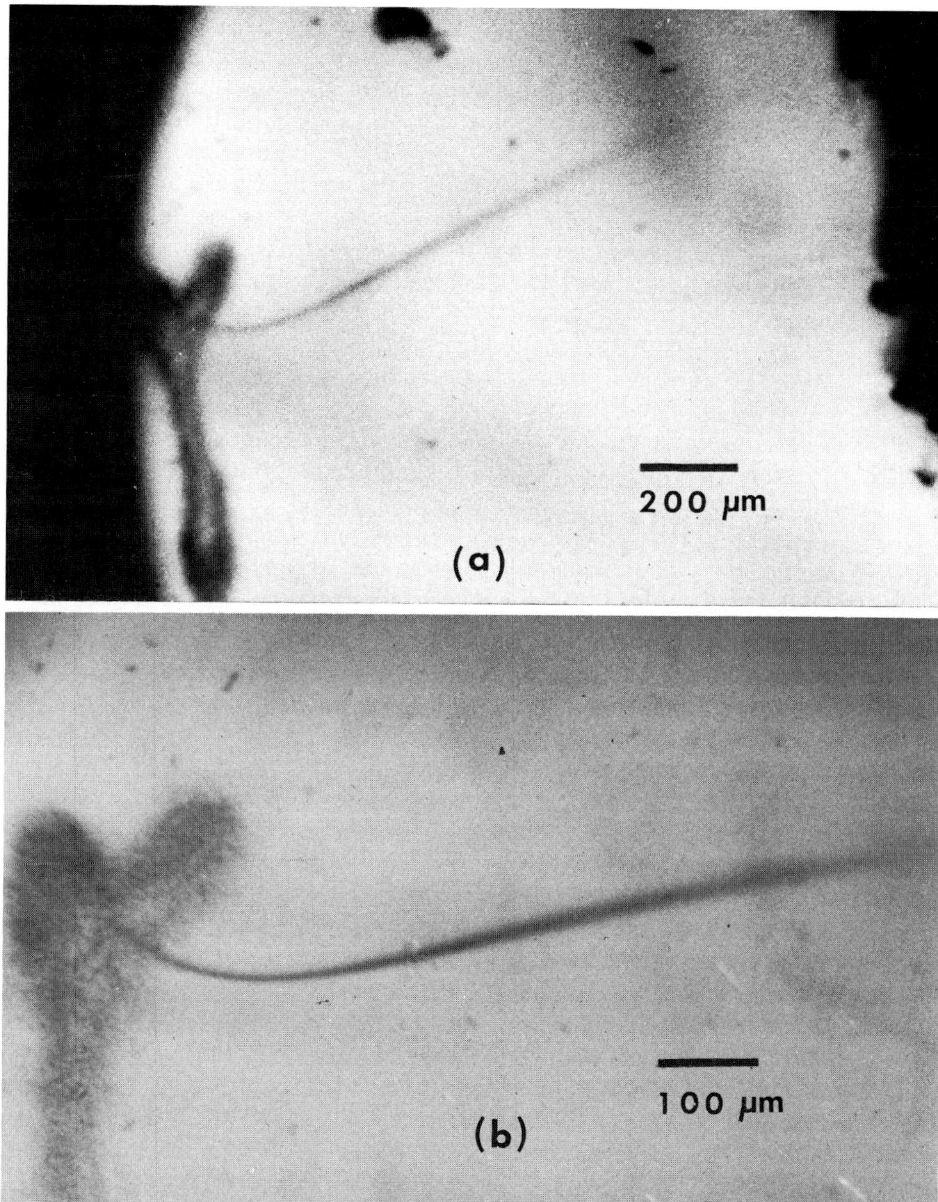

FIG. 5. Photomicrographs of frog lens fiber filled with Procion scarlet. (a) Fiber running from posterior suture toward equator. (b) Fiber curling to enter suture region. (From Rae and Blankenship, 1973.)

driven by an external stepping motor. They reported that if such PD profile measurements were made in the region of the posterior suture, one could frequently measure transient depolarizations of 40–45 mV which quickly returned to the usual resting PD of the lens. They speculated that these PD jumps must be due to the electrode transiently penetrating the extracellular space, showing that at least some fibers maintained a PD across their membranes. Rae (1973b) initially spoke against substantial fiber-to-fiber coupling in the lens. When he placed electrodes into two cells which were very close together, passed 50 nA of current through one, and recorded the induced PD with the other, only a 2.5 mV change was noted. This change was small in comparison to that noted in the linear cables of epithelial cells (Lowenstein, 1966). In retrospect, the experiments actually suggested extensive fiber-to-fiber coupling when one considers the result in light of the cable properties of a spherical volume conductor (Eisenberg and Johnson, 1970). In such a volume conductor, one would expect the induced voltage to fall precipitously as the electrodes are separated.

Rae (1973a) and Duncan (1973, 1974) both reported that the existing data could be explained if the lens fiber membranes were not degenerate, but were electrically coupled through low resistance junctions. They arrived at these conclusions independently. Cohen had previously suggested electrical coupling on the basis of the morphologic findings in 1965. If the cells were electrically coupled, then a PD would exist across the outside layer, but could also exist across any fiber membrane that had a high enough resistance to sustain it. Depolarizing any part of the lens would cause other parts of it to also depolarize because electrical equilibration could occur through the cell-to-cell junctions. Since the resistance of the junctional membrane would be considerably less than the resistance of usual cell membrane, Duncan and Andree's findings that the resistance of the lens appeared to occur entirely across outside layers would then be expected. Duncan's PD and resistance findings did not prove that the membranes of fiber cells were of low resistance, only that the cells were electrically coupled.

Low-resistance junctional membrane in other tissue allows cell-to-cell movement of low-molecular-weight substances other than Na^+, K^+ , and Cl^-, i.e. Procion yellow and fluorescein. Therefore, Rae (1974a) and more recently Rae and Stacey (1976) injected Procion yellow into frog lenses and rat lenses respectively using glass microelectrodes and iontophoretic techniques. They prepared the lenses for histology and looked at sections under a fluorescence microscope. They found that the Procion did not stay within a single fiber; rather it moved readily from one fiber to another without entering the extracellular space. In the frog lens, Rae found that the Procion would not enter the cell from the extracellular space, an observation which has now been consistently verified in rat lenses (Rae, unpublished observations). The conclusion was then drawn that the injected fibers did have intact membranes since they retained the Procion, but that Procion was able to move from one cell to another presumably through the low-resistance junctions. In an attempt to further verify that PDs did exist across individual fibers, Rae (1974b, 1974c) made additional high-resolution PD profile measurements in the frog lens. The measurements were based on the assumption that as the electrode left one cell it would transiently enter the extracellular space before penetrating the next cell. Some, although it be small, PD blip should occur. The spacing of the potential blips should correlate with the histologically measured thickness of lens fibers. When the experiments were actually performed, such correlation was found at least for depths up to 800 μm below the surface. On numerous occasions the electrode tip entered a low-voltage compartment similar to that reported by Rae and Blankenship (1973) earlier. The identity of this low-voltage compartment is still a mystery. Rae (1974c), using junction potential techniques, concluded that it was the extracellular compartment based on its apparent lack of immobile anions. This conclusion is still questionable since the compartment maintained a 30 mV potential, negative with respect to the bathing solution. Again, it appeared that many internal fibers maintained a voltage across themselves. The gross appearance of the resistance and PD occurring only across the outside membranes seemed due to the fact that the cells were electrically coupled. The most recent data dealing with this issue were published by Eisenberg and Rae (1976). In their study, it was surmised that if the internal fibers were extensively electrically coupled, the current–voltage relationships in the lens should look similar to that for a giant spherical cell, i.e., when a current step is passed, the induced voltage response should have two components. One should be a very rapid component that is entirely dependent upon the internal resistivity (R_i) of the lens and the second should be a long-charging component depending only on the "membrane" resistance (R_m) of the lens (Peskoff and Eisenberg, 1975; Peskoff and Ramirez, 1975). Further, the rapid component should show spatial dependence, i.e., be dependent upon the separation of the current passing and voltage measuring electrodes. It should be large when the electrodes are very close together and should decrease precipitously as the electrodes are separated. The long-charging component, however, should be isopotential, independent of electrode separation. When the current–voltage relationships were actually determined for the frog lens, the predictions from the spherical cell analogy were, in part, upheld. By curve fitting techniques, R_m was found to be approximately 2700 Ω-cm^2, whereas R_i was 625 Ω-cm. The value for R_m was similar to that reported previously by

Duncan (1969b) using a similar technique. R_i in comparison to other cells (Katz, 1966), was found to be highly elevated. Pauly and Schwan (1964), using high-frequency bridge techniques, calculated the resistivity of bovine cortical fiber cytoplasm to be 130–170 Ω-cm. Rae and Germer (1974), using a junction potential analysis, found it to be 40–70 Ω-cm in the frog lens. Both values were far lower than that found by Eisenberg and Rae. The authors suggested this high resistivity value occurred because the measurement included the series resistivity of both cytoplasmic and junctional components, with considerably weighting in the direction of the junctional resistivity. More importantly, they reported on the basis of an analysis of the induced PD time constant, that they must be measuring at least 200–300 cm^2 of membrane area, an area far greater than that of the outside surface area of the lens. This analysis was based on the time constant expected from a single parallel R–C circuit, TC $= R_m C_m$. A similar finding can be noted from Duncan's (1969b) work. Although he showed no original records, he reported a requirement of a second or two for the induced PD to plateau. The time constant of approximately 300–500 msec that one would calculate from this statement, his 2200 Ω-cm^2 resistance measurement, and the assumption of a 1 $\mu F/cm^2$ capacitance, would also suggest that in Duncan's measurements, more than an outside layer of the lens was being charged. This remains a point to be addressed more completely. The problem was further complicated by Eisenberg and Rae's report that the long time-constant component of the induced PD was not a single exponential. There were at least two exponentials, probably three, and perhaps a great number more. These findings make the spherical cell analogy inadequate for the lens without major modification.

Other important studies suggest that at least the surface cortical fibers have intact membranes. The optical studies of Bettelheim and his co-workers have previously been described. Kaiser and Maurice (1964) either placed fluorescein on a small spot of bare bovine lens cortex or injected it into a rabbit eye. They found the dye to diffuse readily in the cytoplasm of lens fibers, but to pass slowly across fiber membranes. Maurice and Paterson (1968) and Paterson and Maurice (1971) placed small spots of isotope on bare bovine lens cortex. They showed that the diffusion pattern for ^{22}Na was consistent with extracellular diffusion, the fiber membranes apparently offering appreciable resistance to such diffusion. ^{42}K diffusion appeared to occur largely within cells. The authors suggested that cortical fiber membranes were intact.

The fact that individual lens cells appear to have intact membranes and maintain a PD across their membranes, does not ensure they possess ion pumps for creating gradients. It only means they have adequate resistance to maintain gradients once they are established. Also, none of the preceding studies deal with cells deep in the lens. It is still entirely possible that cells in the deep cortex and nucleus do not have high-resistance membranes. Surely the morphologic data previously stated would suggest some degeneration in membranes deep in the lens. However, in light of the fixation problems noted, one cannot have confidence in these data. Pauly and Schwan (1964), using high-frequency bridge techniques, measured the dielectric constant and electrical conductivity of various regions of the bovine lens. Their results were consistent with the cortical fiber membranes being intact and possessing a capacitance of approximately 1 $\mu F/cm^2$. Results from the deep cortex and nucleus suggested that the fibers there might be degenerate since they provided results similar to those found after treating surface cortical fibers with digitonin, a membrane-disruptive agent. These are interesting measurements that have been largely overlooked. A productive area for further investigation would be that involving the physiologic state of fibers deep in the lens.

B. The Lens as an Epithelial Membrane

Electrophysiologically, the lens is further complicated by the asymmetric location of its ion pumps (Becker and Cotlier, 1962; Kinsey and Reddy, 1965). When Candia *et al.* (1970, 1971) placed an amphibian lens in a modified Ussing chamber, it was apparent that the PD of the lens was also asymmetrically distributed under these conditions (Fig. 6). Similar results were reported by Sperelakis and Potts (1959) and Kinsey and McGrady (1971). It had been a consistent observation that the lens was isopotential when both surfaces were electrically shorted through the bathing medium (Brindley, 1956; Duncan, 1969a; Rae *et al.*, 1970; Paterson *et al.*, 1974; Murata and Taura, 1973; Murata *et al.*, 1974). In Candia's experiments, the circuit was opened between the anterior and posterior, allowing the asymmetric PD to be established and demonstrated. An 80 μm tip Ringer–agar filled electrode was placed through the equator into the center of the lens in order to simultaneously measure the PD across both the anterior and posterior surfaces. The electrode was referred either to the anterior or posterior bath. Routinely, there was a PD of approximately 60 mV, inside negative, across the front surface, and of 32 mV, inside negative, across the posterior surface, with a resultant 28 mV translens PD, anterior surface positive. The short-circuit current was found to be about 30 $\mu A/cm^2$. The importance of the short-circuit current is as follows: ions can move across a membrane in two ways. One way is passively, either in response to a difference in concentration across the membrane, or for a charged substance, in response to a voltage difference across the membrane. The second is by an active process requiring metabolic energy. If

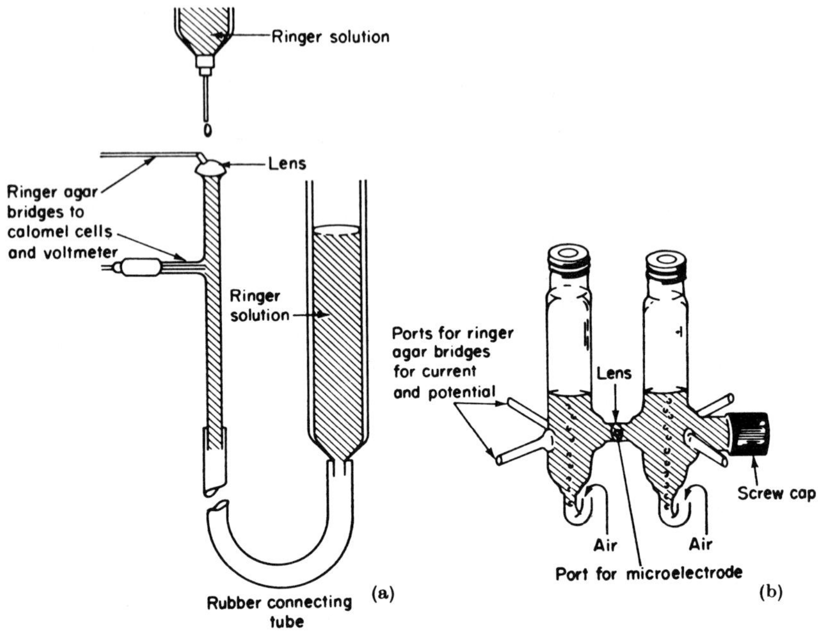

FIG. 6. Chambers used to isolate anterior and posterior sides of toad lens. (From Candia *et al.*, 1970.)

one creates a situation where there is no electrical or chemical gradient across a membrane, any net movements of substances that occur under these circumstances must be actively produced. In measuring the short-circuit current of the lens, identical solutions are placed on both sides to ensure that there will be no translens ionic-concentration difference. Current is passed until the PD across the lens is zero. The electrical current needed to reduce the translens PD to zero when there is no ion concentration difference between the two sides of the lens should be equal to the current produced by active transport processes. To determine which substances are contributing to this short-circuit current and are thus actively transported, it is necessary to measure the unidirectional fluxes in both directions for substances of interest. When this was done (Candia *et al.*, 1970, 1971), the unidirectional fluxes for both K^+ and Cl^- were found to be equal in both directions, whereas there was a net flux of Na^+ from the posterior of the lens to the anterior. This current due to Na^+ amounted to only 6–7 $\mu A/cm^2$, or roughly 25% of the total short-circuit current found. Removal of bicarbonate from both solutions did not affect the short-circuit current as long as the pH remained constant. Therefore, the authors were unable to account for 75% of the short-circuit current. In measuring the unidirectional fluxes, these investigators found they had to wait 40–65 minutes after adding isotope at either surface before a steady flux of the isotope appeared at the opposite surface. This time was presumably necessary for the isotope to mix within the lens and traverse from one side to the other. It was much longer than that found in usual epithelial membrane preparations, but was expected because of the thickness of the lens. According to Bentley (1973), the lenses used were about 50 times thicker than toad bladder.

In Candia's experiments (Candia *et al.*, 1970, 1971), it was possible to pass current across the lens and measure a change in the translens PD that was induced by such current flow. From this, they were able to calculate a translens resistance of 867 Ω-cm^2. They found that the resistance of the posterior surface was always approximately 1.5 times that of the anterior surface. The actual posterior resistance was 464 Ω-cm^2 versus 308 Ω-cm^2 for the anterior surface. They have, to date, made no attempt to identify the specific structures giving rise to these resistances. Importantly, they found that the anterior surface of the lens was significantly depolarized by increasing the K^+ concentration in its bathing solution, whereas increasing the K^+ concentration at the posterior surface of the lens had very little effect on the PD there (Table I). These results suggest that the anterior surface of the lens has a greater permeability to K^+ than the posterior surface. No K^+ efflux studies across the isolated surfaces have yet been done to confirm these results. The removal of Cl^- from both surfaces had very little effect on either PD, suggesting a passive chloride distribution. As expected from Goldman equation considerations, the removal of the Na^+ from the posterior surface caused the posterior PD to slightly hyperpolarize. Removal of Na^+ from the anterior surface caused a significant depolarization there. Candia *et al.* (1970, 1971) interpreted this result to mean that some anterior metabolic process was inhibited, whereas Duncan (1974) suggested that the anterior surface must have been more permeable to the Na^+ substitute choline$^+$ than it was to Na^+. This Na^+-free finding at the anterior surface is certainly unexpected and requires further investigation. At the very least, the results of these investigations clearly establish that the electrical properties of the lens anterior and posterior surfaces are very different. It seems certain that there is an active mechanism that pumps Na^+ out through the anterior surface. It is not at all clear why such a pump mechanism might exist in the lens. It would appear better to have a Na^+ pump located entirely around the surface of the lens actively extruding Na^+ from inside the lens to the outside, thus aiding in the maintenance of

TABLE 1

Effects of Ionic Changes on Total PD and across Anterior and Posterior Sides of Toad Lens

Bathing solution	Potential difference, mV		
	Posterior[a]	Anterior[a]	Total
		Normal	
	31	57	27
		Change in PD +	
Cl-free posterior side	0	0	+2
Cl-free anterior side	0	+4	
Na-free posterior side	+6	0	−12
Na-free anterior side	0	−7	
K 75, Na 31, Cl 33 mEq/liter, posterior side	−6	0	−27
K 75, Na 31, Cl 33 mEq/liter, anterior side	−7	−34	
K-free posterior side	+2	0	+5
K-free anterior side	0	+7	

[a] Intralenticular electrode taken as reference for anterior and posterior PD. Posterior solution electrode taken as reference for total PD. (+) Positive signs indicate increment in PD and (−) negative signs indicate decline in PD.

lenticular water balance (Duncan and Croghan, 1969). However, to date, no Na^+ pump has been found associated with the posterior surface. Rather, it appears to exist only on the anterior surface, presumably associated with the epithelial layer of cells located there. Ion pumping by anterior fiber cells has not been unequivocally disproved.

Additional characterization of this asymmetric, ion-pumping mechanism was provided by Bentley and Candia (1971) when they investigated the effect of metabolic inhibitors on the short-circuit current of the Ussing chamber-mounted toad lens. They found that the only inhibitors that appreciably affected the short-circuit current were iodoacetate (IAA) and ouabain. Dinitrophenol, ethacrynic acid, carbonic anhydrase, and amiloride had very little effect. The action of IAA was much faster than that of ouabain, a result which they said was expected because ouabain-inhibitable sodium transport is only 25–30% of the total short-circuit current. They concluded that the mechanism for providing the short-circuit current was dependent on anaerobic rather than aerobic metabolism. Recently, Amphotericin B (Bentley and Candia, 1975), hyperosmotic agents (Yorio and Bentley, 1976), and heavy metals (Bentley and Yorio, 1976) were all shown to alter the short-circuit current of the toad lens when they were placed in the anterior bathing solution. These agents were ineffective in altering the short-circuit current when they were placed in the posterior bathing solutions. These results again strongly implicate the anterior epithelial cells in ion active transport.

Candia (1973) dealt with the discrepancy between the total short-circuit current and that produced by transport of Na^+ ion. He noted that previous work by Toyofuku and Bentley (1970) had shown toad lenses incubated *in vitro* have a steady, but slow loss of ions over a prolonged time period. In their contribution, they noted that 1.5 times as much Na^+ was gained by the toad lens as K^+ was lost. Candia reasoned as follows: The anterior lens permeability to K^+ is greater than that of the posterior (Candia *et al.*, 1970, 1971). Conversely, the posterior lens permeability to Na^+ is greater than that of the anterior (Kinsey and Reddy, 1965; Paterson, 1969b). Therefore, when the ion gradients run downhill, a net short-circuit current could result since anterior K^+ loss would exceed its posterior loss the same time that posterior Na^+ gain exceeded Na^+ gain from the anterior. When Candia (1973) simultaneously measured the K^+ loss and the Na^+ gain from toad lenses, he found that the total number of microequivalents moved per hour per area could account for the difference between the net Na^+ flux and the short-circuit current measured previously. He therefore speculated that the total short-circuit current was due to active anterior Na^+ transport moving Na^+ from the posterior of the lens to the anterior with additional simultaneous current being produced by the ion gradients in the lens running downhill. It is not clear how this explanation correlates with the previous observation that IAA reduced the short-circuit current more and faster than ouabain. It would seem that the effect of IAA would have to be, at least in part, to cause the passive ion gradients to run downhill faster.

Candia (1973) also noted that the Na^+ transport in the amphibian lens was only about 25% of that reported in other amphibian tissues. He noted that this low value might be in error because of the way in which the short-circuit current was measured. When one short-circuits the entire translens PD, the PD between the inside microelectrode and the anterior surface is not short circuited. Even if one increases the current so that this PD is forced to zero, an ion gradient still exists between the inside of the lens and the anterior bathing solution. In an attempt to more adequately estimate the short-circuit current produced by the anterior surface, Candia replaced the anterior bathing solution with one whose composition was similar to the ion composition of the lens as determined by flame photometry. Under these conditions, a short-circuit current of 60 $\mu A/cm^2$ was found when the anterior PD alone was short-circuited. Candia believes this to be the true short-circuit current of which the anterior Na^+ pump is capable. Unfortunately, it is not possible under these circumstances to measure the unidirectional Na^+ fluxes to verify his hypothesis.

There can be little doubt that the isolated toad lens functions much the same as other epithelial membranes in that it has an asymmetrically located Na^+ pump that produces a sizeable short-circuit current. No entirely convincing evidence can be found for active K^+ pumping (Candia, 1973). Fowlks (1973) has demonstrated a net lenticular H_2O flux from posterior to anterior. In comparing the toad lens to other membranes it would appear that it is a relatively leaky membrane. The translens resistance in the toad is almost 1000 Ω-cm^2, a value quite similar to bladder, skin, and stomach, and much higher than gallbladder, intestine, and renal tubule (Bentley, 1973). It is probable, however, that this relatively high resistance is much more related to its thickness than its tightness. The 308 Ω-cm^2 and 464 Ω-cm^2 values for the anterior and posterior resistances, respectively, are consistent with a moderately leaky epithelial tissue. We are not here referring to leaks around the clamped preparation. Candia *et al.* (1971) showed clearly that the unidirectional water flux in the toad lens was only about 1/150th of that in the toad bladder, a result that could not have been obtained if there had been a leak around the preparation. Candia also calculated that the diffusion coefficient for water in the lens was $5.5 \times 10^{-7} cm^2/sec$, which is virtually identical to the value found in toad bladder (quoted in Bentley, 1973). Therefore, the slow water movement in the lens is not due to any particular way that water is handled there, rather it must be due to the thickness of the lens.

C. Resistance Measurements

Electrophysiologists have always had an interest in measuring the resistance of tissues to electrical current flow. Since it is well known that such current is carried by the charged ions in solution, measurement of tissue resistance is therefore an indirect measurement of the impediment offered by the tissue to the movement of ions. It has been attractive to model tissues as though they were circuits composed of resistors, capacitors, and inductors. Such an approach has been fruitful in dealing with electrically excitable tissues. It must be remembered, however, that since tissues are not actually composed of such circuit elements, and since resistance is entirely a calculated phenomenon, the resistance measurements must be interpreted in light of some model. The information and interpretations gained from the measurement are only as valid as the degree to which the model fits the actual situation. Even if a model seems to predict the experimental observations, one cannot always be certain that some entirely different circuit model might not predict just as well. Therefore, one must be cautious in interpreting the results from resistance measurements. Nonetheless, such electrical approaches are of great interest since electrical techniques can be very precise and versatile, allowing one to deal with questions that might not be dealt with otherwise.

Resistance measurements on epithelial-type tissues clamped in Ussing chambers have recently been approached sophisticatedly (Frömter, 1972; Frömter and Diamond, 1972). Since this approach is important to understanding the results of lens resistance measurements to date, the approach will be briefly considered. The basic question is as follows: When current is passed across an epithelial tissue, do the ions carrying the current go between the epithelial cells, through the epithelial cells, or both? If both pathways are used, as is certainly the most reasonable initial assumption in any tissue, what is the percentage of current carried by each pathway? It is also of great interest to know if the resistance of the through-the-cell pathway is equally distributed between both surfaces. The answers to these questions are not easily obtained and are certainly not obtainable from measurements made to date in the crystalline lens of the eye.

Figure 7a shows a circuit diagram which might fairly accurately represent the epithelium of a simple epithelial tissue (Frömter, 1972; Frömter and Diamond, 1972). R_m represents the resistance of the cell membrane on one side, whereas R_{bb} represents the resistance of the the cell membrane

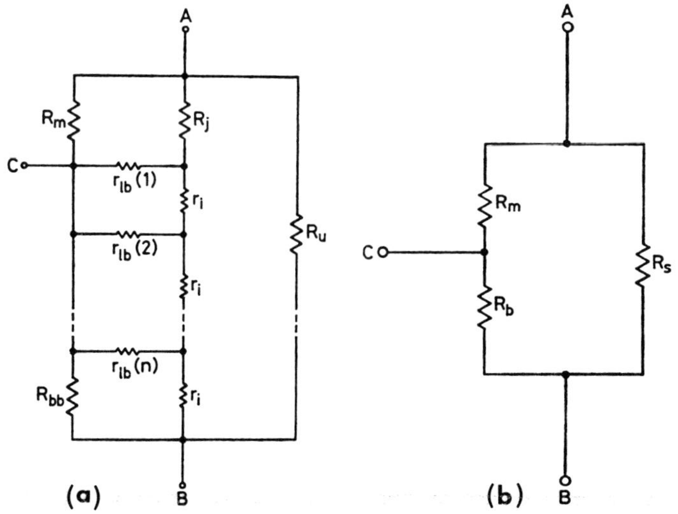

FIG. 7. (a) Equivalent circuit of epithelial cell with paracellular shunt path: R_m and R_{bb} denote luminal and basal cell membrane resistances, respectively; R_j and R_u are resistances of junctional complex and unspecified shunt path, respectively; r_i is resistance of lateral extracellular space; r_{lb} is resistance of lateral part of basal cell membrane. A, B, and C are the luminal, interstitial, and intracellular compartments, respectively. (b) Lumped parameter circuit, determined in the experiment. (From Frömter, 1972.)

on the opposite side. R_j and R_u are the resistances of the junctional complexes (around cells, not cell to cell) and an unspecified shunt path (i.e., around the preparation), respectively. The resistances of the lateral extracellular space r_i and the lateral part of the cell membrane r_{lb} form a cablelike element consisting of n subunits. A, B, and C are the luminal, interstitial, and intracellular fluid compartments, respectively, into which electrodes may be placed. Note that cytoplasmic resistivity has been neglected. This circuit cannot be accurately analyzed directly since only points A, B, and C are accessible to electrodes. Only three lumped resistances can be obtained from such a circuit. They are shown in Fig. 7b.

It is usually assumed that r_i (resistance of extracellular fluid) of Fig. 7a is negligible in comparison to both r_{lb} and R_{bb}. Therefore, r_{lb} and R_{bb} form a parallel resistance network where $1/R_b \sim 1/R_{bb} + n/r_{lb}$. R_s from Fig. 7a would be made up of R_j and R_u in parallel. Thus $1/R_s \sim 1/R_j + 1/R_u$. Duncan (1973, 1974) has used a similar circuit to analyze Candia's *et al.* (1970, 1971) resistance data and to make predictions of potentials which would exist inside the lens when it was immersed in a bathing solution. In his model, he apparently assumes R_s is infinite, an assumption for which the justification is unclear. Candia *et al.* (1970, 1971) wisely avoided an analysis of the resistances which they measured. If one were to try to analyze the lens resistance to determine the real resistance of anterior and posterior lens "membranes," one might define the anterior membrane resistance to be analogous to R_m, and the posterior to R_{bb}. The resistivity r_i of the extracellular channel, might be assumed to be negligible as per Frömter's analysis, but it should be noted that since the extracellular channels of the lens are long, narrow, and tortuous, such an assumption might not be acceptable. If not, then R_b would also include a contribution from r_i and would indeed be a complex resistance. To date, even the simplified circuit in Fig. 7b has not been determined for the lens. In order to analyze such a circuit, three measurements are necessary. First, passing current from A to B while recording the voltage drop from A to B would give an estimate of the total resistance $[1/R_t = 1/R_s + 1/(R_m + R_b)]$. This measurement has been made by Candia *et al.* (1971). Second, passing current from A to B while recording the voltage from both C (microelectrode) to A or C to B would give $\Delta V_m/\Delta V_b = R_m/R_b = A$. A is called the voltage divider ratio. Note that A does not give the ratio of resistances across the anterior and posterior membranes because R_b is composed of contributions from both posterior membrane resistance and the resistance of the lateral membrane and possibly that of the extracellular fluid. This measurement was also made by Candia *et al.* (1971) although one cannot be certain their measurement is entirely intracellular because of the large microelectrode used. Third, current is passed from the microelectrode at C into both compartments A and B while recording the induced voltage between C and A and C and B. The circuit reduces to that shown in Fig. 8. Then $1/R_z = 1/R_m + 1/R_b$. From the results of these three methods, one should in principle be able to calculate R_m, R_b, and R_s. Unfortunately, measuring R_z is usually much more complicated than described. When adjacent cells are electrically coupled, as they are in most epithelial cells and in the lens, current does not leave the tissue via R_m and R_b alone. Rather, it also flows into adjacent cells, presumably through gap junctions. In this case, it is no longer possible to derive R_z from the voltage drop across R_m and R_b. Instead the lateral voltage spread must be determined and R_z must be calculated taking the cable properties of the tissue into consideration. Whereas it is possible to derive an equation to describe the cable behavior of a single layer of coupled epithelial cells in a flat sheet (Frömter, 1972), at present we know of no satisfactory equation to describe the cable properties of a sphere of coupled cells, the apparent case for the lens. Until such a relationship is available, it is impossible in principle to determine the relative resistances for the anterior and posterior lens membranes. The problem is actually much worse when one considers that the majority of our discussion to this point has assumed the lens to be a single layer of flat epithelial cells. Actually, one minimally would have to consider the circuit diagram for the lens seen in Fig. 9. Q is a network similar to the one just described, but represents the anterior epithelial layer only. R_i is the extracellular resistance, R_c the gap junctional resistance, R_{Fm} the fiber membrane resistance, C_{Fm} the fiber membrane capacitance, and R_{ps} the shunt resistance around the entire preparation. R_{ps} is presumably due to either the bathing medium and capsule if the lens is totally immersed in a bath, or the leak around the preparation if the lens is in an Ussing

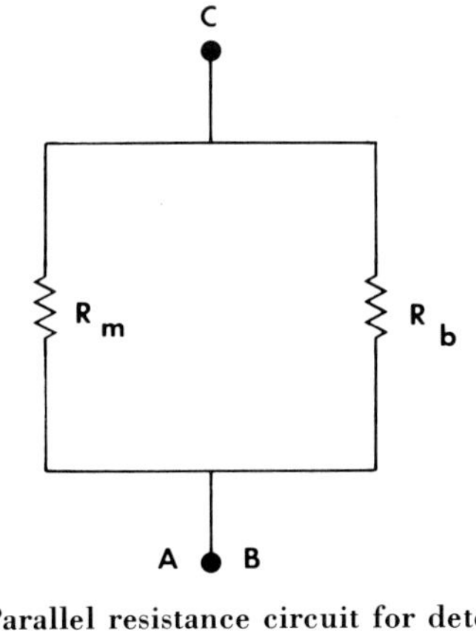

FIG. 8. Parallel resistance circuit for determining R_z.

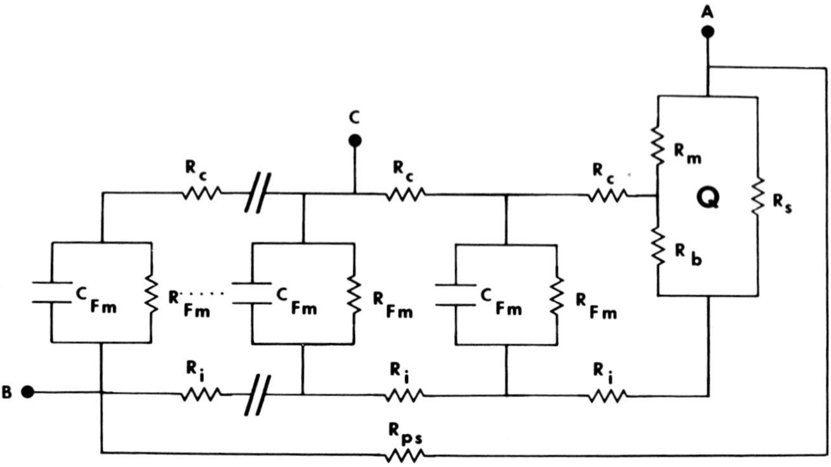

FIG. 9. Equivalent circuit for lens. Q represents anterior epithelium. R_c, R_i, and R_{Fm} are coupling resistance, extracellular resistance, and fiber membrane resistance, respectively. R_{ps} and C_{Fm} are the shunt resistance around the lens and fiber membrane capacitance, respectively. Circuit is broken to denote repeating cable of unknown length.

chamber. This circuit depicts the epithelium attached to a long, repeating subunit cable of unknown length that is composed of lens fibers. The fibers making up the posterior surface are simply part of the cable. One should remember that although the data of Candia *et al.* (1971) make it appear that the posterior side of the lens has a higher resistance than the anterior side, it is impossible with our present understanding of the lens to tell if the difference is really a membrane-resistance difference or different cable and shunt values associated with the two sides. It is not justifiable to ascribe different ionic conductances to the anterior and posterior surface membranes based on the resistance measurements to date.

It is possible to approach the resistance of the lens from a completely different circuit model. That is, the lens can be assumed to be a giant spherical cell with membrane resistance located only at the outside of the sphere. Such an approach has been used by Duncan (1969b), and Eisenberg and Rae (1976). The lens is assumed to be adequately represented by a single parallel R–C circuit. When a rectangular, constant-current pulse is applied to this circuit, the voltage across the capacitor exponentially rises until the capacitor is charged. The voltage across the capacitor when it is fully charged can be used to calculate the membrane resistance. For a spherical cell, $\Delta V = \Delta I R_m / \pi a^2$, where ΔV is in volts, ΔI is in amps, R_m is in Ω-cm^2, a = radius is in cm. Using basically this analysis, both Duncan (1969b), and Eisenberg and Rae (1976) obtained values for R_m of about

2200–3000 Ω-cm^2. Such values are based on an entirely different circuit model than the translens resistance measurements of Sperelakis and Potts (1959) and Candia *et al.* (1971). There is no reason to believe that their translens measurements should give the same resistance values as the R_m determined using the spherical cell model. The fact that the resistance values of Sperelakis and Potts (1959) are about the same as those from Duncan (1969b), and Eisenberg and Rae (1976) should be fortuitous because of the different resistance models involved. The discrepancy between the approximately 2000 Ω-cm^2 resistance of the bovine lens (Sperelakis and Potts, 1959) and the 1000 Ω-cm^2 resistance of toad lenses (Candia *et al.*, 1971) presumably can be explained by the difference in thickness of the two lenses. At the moment, the lens-resistance problem cannot be analyzed further without a theoretical description of the cable properties expected from a spherical mass of coupled fiber cells. Both the epithelial-resistance model and the spherical-cell model cannot properly determine lens membrane resistance properties at this point because both require a cable analysis for which, at the moment, there is no adequate theoretical treatment. It would appear that resistance measurements in the lens are a fruitful area for future endeavor.

D. Ion Distribution and Movement

1. Introduction

There are a number of previous reviews dealing with ion distribution and movement in the lens (Van Heyningen, 1969; Kuck, 1970; Paterson, 1972; Duncan, 1973, 1974). An excellent bibliography on the subject is provided by Paterson (1972).

It is well known that the concentrations of the major ions, Na^+, K^+, and Cl^-, in the lens are very different than their concentrations in either aqueous or vitreous humors (McKay *et al.*, 1932; Fisher, 1933; Salit, 1943; Harris and Gehrsitz, 1951; Heinrichs and Harris, 1956, 1957; Andree, 1958b; Amoore *et al.*, 1959; Merola *et al.*, 1960; Harris and Gruber, 1962; Hart *et al.*, 1963; Thoft and Kinoshita, 1965a,c; Harris and Becker, 1965; Harris, 1966; Aviram *et al.*, 1966; Matuk *et al.*, 1969; Paterson, 1969a, 1970a, 1970b, 1973; Duncan, 1969a; Candia *et al.*, 1970; Reddy, 1972). Table II shows results from rabbits and toads for Na^+, K^+, and Cl^- concentrations of the aqueous humor, vitreous body, and crystalline lens. The lens shows a markedly elevated K^+ concentration along with markedly reduced Na^+ and Cl^- concentrations in comparison to either aqueous or vitreous. These values are similar to those expected when one compares an intracellular environment to an extracellular environment.

TABLE II

Ion Concentration Data for Rabbit and Toad[a,b]

	Na	K	Cl		Na	K	Cl
Rabbit				Toad			
Lens	16	135	16	Lens	17	112	17
Aqueous humor	140	5.0	105	Aqueous humor	119	4	109
Vitreous humor	140	5.2	105	Vitreous humor	119	4	109

[a] Rabbit data from Paterson (1972). Toad data from Duncan (1969a).
[b] Concentrations are millimoles per kilogram of H_2O.

2. Ion Distributions within Lens

The preceding ion concentration results come from experiments in which the entire lens is ground up or solubilized and the ion concentrations measured from the resulting slurry or solution. Such procedures cannot discern an asymmetry in distribution of the ions within the lens. Such asymmetries must be demonstrated by separating the lens into layers or sections and carefully analyzing the ionic constituents of each of the divisions. Amoore *et al.* (1958) separated the bovine lens into a number of concentric layers and analyzed for Na^+ and K^+ in each. They found that Na^+ was most concentrated near the surface of the lens, becoming less concentrated as one moved toward the nucleus. A similar result was noted for K^+, although the gradient was less severe. There was a dynamic component to each gradient since the gradients became less severe as the lens was cooled. They speculated that the gradients were in part produced by ion "pumping" activity in the lens fibers. Phillipson (1969a,b) and Bando *et al.* (1976) found an opposite gradient for lens proteins, i.e., nucleus more concentrated than cortex. Becker and Cotlier (1962) and Kinsey and Reddy (1965) each found that when the rabbit lens was incubated in solutions containing ^{86}Rb, a concentration gradient for the labeled ion was quickly created wherein the Rb^+ concentration was higher on the anterior side than on the posterior side. Paterson (1969a, 1973) showed that an ionic gradient existed from the anterior to the posterior in the normal, unincubated bovine lens. His experiments were similar to those of Becker and Cotlier (1962) in that he rapidly froze the lens, trephined a core from front to back, cut the core into small pieces, and assayed for Na^+ and K^+ in each of the pieces. He found that in normal lenses (without isotopes), there was a slightly higher K^+ concentration on the anterior side than on the posterior, whereas Na^+ was more concentrated posteriorly than anteriorly. A similar gradient for K^+ was recently demonstrated by Horowitz *et al.* (1976) from proton microprobe studies. Paterson hypothesized that the lens has two sets of superimposed gradients; the first, the surface to depth gradient reported by Amoore *et al.* (1958); the second, the anterior to posterior ion gradient just described. These studies clearly showed that the ion distribution within the lens was neither uniform nor simple.

3. Active Ion Transport

Since the ion gradients for Na^+, K^+, and Cl^- that exist between the inside of the lens and the bathing solution are now well established, it is of interest to consider how these gradients are maintained. This question is particularly important in the lens that is only 60–70% H_2O but 30–40% protein. If the lens were allowed to reach Donnan equilibrium, the fixed negative charges of the highly concentrated protein would cause considerable elevation of lens Na^+ and H_2O (Duncan and Bushell, 1975, 1976). See Duncan and Croghan (1969) for a complete discussion of this topic. A Na^+ extrusion mechanism would thus seem of great benefit to the lens. It has been known for some time that there is an active component involved in lens ion gradient maintenance. Harris and Becker (1965) have adequately reviewed the early literature involving such active ion transport in the lens. In a whole series of early reports, Harris and others (Harris and Gehrsitz, 1951; Harris and Nordquist, 1954; Heinrichs and Harris, 1957; Harris *et al.*, 1959; Kinoshita *et al.*, 1961; Kern *et al.*, 1962; Harris and Gruber, 1962; Harris and Becker, 1965; Harris, 1966) showed that when the mammalian lens was cooled, it accumulated Na^+ while simultaneously losing K^+. When the temperature of the lens-incubating solution was reestablished at 37°C, there was a gradual accumulation of K^+ inside the lens against a concentration gradient, while there was a concomitant reduction in the lens Na^+, also against a concentration gradient. This reestablishment of the gradients after cold induction was found to be blocked by a variety of metabolic inhibitors and other drugs. Though more specific proof will be given later, there is little doubt that active ion transport does occur in the lens. The basic questions are: What is the site and/or sites of the active transport processes and which of the ions are actively transported?

4. Locations of Transport Mechanisms

The first of these questions has been actively studied by a number of investigators (Becker and Cotlier, 1962; Garcia *et al.*, 1964; Kinsey and Reddy, 1965; Bonting, 1965; Bonting *et al.*, 1963; Paterson, 1969a, 1973; Candia *et al.*, 1970, 1971; Riley, 1970; Bentley and Candia, 1973; Yuge *et al.*, 1973a,b,c; Palva and Palkama, 1974a,b, 1976; Horowitz *et al.*, 1976). Becker and Cotlier (1962), as previously mentioned, incubated rat lenses in solutions containing ^{86}Rb. After a 6- to 12-hour incubation, the lens was quickly frozen. A core from the center was then trephined, cut into individual pieces, and the ^{86}Rb in each portion was then counted. Becker and Cotlier found that the lenticular Rb^+ concentration was higher in regions close to the epithelium than it was posteriorly. They suggested that this distribution resulted from active Rb^+ transport into the lens by the epithelium, with little or no transport occurring at the posterior surface. This suggestion was strenthened when they repeated the experiment in a bathing solution containing both Rb^+ and ouabain, an effective inhibitor of the Na^+–K^+-activated ATPase enzyme known to be involved in Na^+ and K^+ transport. Exposure to ouabain resulted in a marked reduction in the accumulation of ^{86}Rb by the anterior surface, a result anticipated if an active "pump" in the epithelial cells was being inhibited by the ouabain. It is of interest to note that ouabain also caused a decrease in Rb^+ uptake at the posterior surface. Duncan (1974) has pointed to this result as indirect evidence of active transport by the posterior fibers of the lens. However, Paterson *et al.* (1975) and Hightower and Kinsey (1977) have recently found that in the rabbit lens there is a 10–15 mV electrogenic potential that is rapidly depolarized by the addition of ouabain to the bathing medium. The reduced posterior uptake of Rb^+ in Becker and Cotlier's experiments may have been due entirely to this reduction in membrane potential with resultant alteration in the driving force moving Rb^+ into the lens. Their experiments, therefore, do not provide unequivocal evidence for active transport by the posterior fibers of the lens. Kinsey and Reddy (1965) also elegantly demonstrated an asymmetry in the pumping ability of the lens when they isolated the two surfaces of the lens in a specially designed apparatus. They were able to selectively subject either the anterior or posterior surfaces to solutions in which isotopes of important ions had been added and to subsequently follow the rate at which these ions appeared in the lens (Table III). They found that K^+ was accumulated about three times faster through the anterior surface than through the posterior surface. Conversely, Na^+ entered the lens four times more readily through the posterior than the anterior. When these experiments were repeated with ouabain in the incubating fluid, the preferential uptake of K^+ on the anterior surface was abolished and roughly equal uptake was noted for the anterior and posterior surfaces. No evidence for a preferential active uptake of chloride was noted at either surface. Lens decapsulation, a procedure that removed both the capsule and anterior epithelium also removed the asymmetry in Na^+ and K^+ uptake between the two sides. The authors then proposed what has become known as their "pump-leak" model (Kinsey, 1965; McLean *et al.*, 1969). They suggested that K^+ ion was actively pumped from the anterior bathing medium into the lens. Once in the lens, the K^+ diffused through the interior of the lens to the posterior surface where it leaked out. Conversely, Na^+ was thought to leak into the

TABLE III

Influx of Ions in Counts per Minute per Lens When a Single Surface Is Incubated for 1 Hour in Media Containing 10^5 Counts per Minute per Milliliter of Radioactivity[a]

Addition to media	Intact (anterior)	Intact (posterior)	Decapsulated (anterior)	Decapsulated (posterior)
Potassium (^{42}K)				
None	3660 ± 740 (21)[b]	1180 ± 290 (25)	1420 ± 400 (9)	1370 ± 320 (11)
Ouabain (10^{-5} *M*)	990 ± 180 (10)	1000 ± 220 (11)		
Sodium (^{22}Na)				
None	162 ± 74 (22)	658 ± 190 (17)	1390 ± 229 (20)	1503 ± 270 (19)
Ouabain (10^{-5} *M*)	296 ± 116 (14)	573 ± 220 (20)		
Chloride (^{36}Cl)				
None	239 ± 108 (59)	370 ± 170 (50)	880 ± 165 (16)	790 ± 134 (13)
Ouabain (10^{-5} *M*)	230 ± 52 (14)	313 ± 131 (11)		

[a] Data from Kinsey and Reddy (1965).

[b] Number of lenses shown in parentheses.

lens across the posterior surface, diffuse through the lens to be actively pumped out through the anterior side. Their experiments also showed that both K^+ and Na^+ leaks could occur across the anterior surface. Paterson (1969a, 1973) was able to verify the previous results of Kinsey and Reddy (1965) when he detected anterior–posterior gradients for Na^+ and K^+. If the system proposed by Kinsey and Reddy was operative, the K^+ concentration should be higher in the anterior part of the lens and lower in the posterior. An opposite gradient should occur for Na^+. Paterson found these predicted gradients. Perhaps the most convincing verification of the Kinsey–Reddy model came from Paterson's (1973) study where he looked at the reestablishment of the gradients after cooling of lenses *in vitro*. He was able to show that on rewarming these cold-depleted lenses, the anterior to posterior gradients were reestablished. However, the change in ion concentrations on the anterior side far preceded that occurring on the posterior side. This result suggested that the active transport was occurring on the anterior surface and that any posterior process was passive. Additional evidence for Kinsey and Reddy's hypothesis was provided by Bonting (1965) and Bonting *et al.* (1963) when they showed that the lens contained Na^+–K^+-activated ATPase known to be involved in active transport of Na^+ and K^+ in other tissues. They found that more than half of the total lens ATPase activity was located in the epithelium, again pointing to the epithelium as the principal site of active transport of ions. More recently, Yuge *et al.* (1973a,b,c) performed autoradiographic studies on the uptake of ^{22}Na and ^{86}Rb by the rabbit lens. They were able to show that in normal lenses, Rb^+ entered the lens far more rapidly in areas near the anterior epithelium, whereas Na^+ entered more rapidly through the posterior surface. When they repeated the experiments with ouabain or IAA in the bathing medium, the asymmetry disappeared and there was roughly equal uptake around the entire surface of the lens. Also, they demonstrated that for short-term incubation of normal lenses in a solution containing ^{86}Rb, an anterior to posterior gradient was established. Finally, Palva and Palkama (1974a,b, 1976) were able to show histochemically that Na^+–K^+ ATPase in the lens was primarily located on the lateral border of anterior epithelial cells. It could not be unequivocally demonstrated in cortical lens fibers. It was not found in appreciable quantities on either the anterior or posterior surface of the epithelial cells. Rather, it was found primarily on the lateral walls. The implications of this lateral location are not yet clear, but represent an interesting area for additional research.

As a result of these many studies, along with the previously discussed electrical studies of Candia *et al.* (1970, 1971), we must conclude that the sites for active transport of ions in the lens are asymmetrically located. It appears that the anterior epithelial cell assumes the primary responsibility for active ion pumping, whereas at the present time, only a passive role can be attributed to the posterior lens fibers or to lens fibers in general. Cotlier and co-workers (Cotlier and Apple, 1973; Cotlier, 1973; Cotlier *et al.*, 1974, 1975) have argued that a pump leak occurs across individual fiber cells based, in part, on the fact that selected antibiotics or surface active agents cause individual lens fibers to hydrate, swell, and rupture at different rates. The result would seem to suggest that functional membranes do exist around these fibers. It is not clear, however, how they arrived at the conclusion that the fibers have pumps. Brief incubation of the posterior lens surface in 1% cetyl pyridinium chloride (CPC) caused no effect or even stimulation of the "Rb^+ pump" whereas incubation of the anterior surface resulted in marked reduction in the Rb pump (Cotlier *et al.*, 1974). One might interpret this result to mean that there is no pump to be destroyed on the posterior surface and that the pump stimulated following posterior CPC incubations was the anterior pump increasing its activity in an attempt to keep up with the greater posterior leak caused by the CPC. It would seem that a posterior fiber pump has still not clearly been demonstrated. The primary evidence against anterior fiber cells being able to pump ions comes from the mechanical lens decapsulation studies of Kinsey and Reddy (1965) where they were unable to demonstrate net accumulation of ^{86}Rb and ^{42}K across the anterior lens surface after decapsulation. As pointed out by Reddy (1973), one cannot rule out that the decapsulation procedure might have damaged the surface fiber membranes. The increased Na^+ entry across both the anterior and posterior lens surfaces (Kinsey and Reddy, 1965) following decapsulation would support that suggestion, particularly when one considers the substantial immediate depolarization that occurs when the rabbit lens is mechanically decapsulated. The usual, approximately -70 mV PD (Brindley, 1956; Paterson and Rae, 1974) falls to about -19 mV (Kinsey and McGrady, 1971). Such a depolarization would reduce, not increase, the driving force tending to move Na^+ into the lens. Because of the large increase in the apparent passive leak of the fiber membranes, it is possible that existing active pumps might not be demonstrable after decapsulation. It seems clear from data previously presented that the posterior fiber membranes do not demonstrate appreciable active ion transport. We cannot be entirely certain about the anterior fiber membranes.

5. Ion Flux Measurements

Additional information on the nature of active transport of ions in the lens has come from experiments where either the uptake or efflux of labeled ions has been investigated. Of course, the three major ions of interest are Na^+, K^+, and Cl^-. Becker (1962), however, found that the rabbit lens handled K^+ and Rb^+ in an almost identical fashion. Therefore, since that time, many experiments have been conducted using ^{86}Rb rather than ^{42}K with the information obtained directly extrapolated to K^+.

a. Potassium and Rubidium

There are a number of questions that one would want answered in order to understand the way in which the lens handles K^+. One would expect the K^+ influx to be both active and passive, whereas the K^+ efflux should be entirely passive. Can these expectations be verified? Are the passive K^+ permeabilities of the anterior and posterior surfaces of the lens identical? There have been a number of studies in which K^+ and Rb^+ fluxes in normal lenses were investigated (Von Sallmann and Locke, 1951; Becker, 1962; Becker and Cotlier, 1962, 1965; Kinsey and McLean, 1964, 1970; Kinsey and Reddy, 1965; Thoft and Kinoshita, 1965b; Cotlier *et al.*, 1968; Duncan, 1969a, 1969c; Paterson, 1969b, 1970c; Candia *et al.*, 1970, 1971). This list is, of course, not complete, but the cited work covers the major findings. In his original investigation, Becker (1962) found that the influx of the ^{86}Rb was inhibited by increasing either the K^+ or the Rb^+ concentration in the bathing solution. He found, using Michaelis–Menten kinetics, that the two ions were competitive inhibitors, suggesting mediated transport. He was also able to show that the Rb^+ influx was markedly reduced by decreased temperature and by metabolic inhibitors, particularly iodoacetate and ouabain. Therefore, he showed accumulation of Rb^+ against a concentration gradient by a mechanism demonstrating temperature dependence, saturation, competitive inhibition, and the influence of metabolic inhibitors. These are criteria that must be satisfied in order to hypothesize active transport. Kinsey and McLean (1964), using similar techniques for K^+ transport, found that the transport system had a K_m or half-saturation value of 0.85 mM and a V_{max} of 1.3 mM per lens per hour. They calculated that for the 5 mM K^+ concentration in the aqueous of the rabbit, the K^+ pump would be operating at about 85% capacity. Kinsey and Reddy (1965), as previously stated, demonstrated that the ^{42}K active transport took place entirely across the anterior surface.

The passive permeability of the lens to ^{86}Rb was determined by Becker and Cotlier (1965) from Rb^+ efflux studies on the rabbit lens. They showed that the Rb^+ efflux was unaltered by enzymatically removing the capsule with collagenase, thus dispelling any notion that the capsule might offer appreciable resistance to the movement of Rb^+ or K^+. Strangely, they found that a number of metabolic inhibitors and steroids increased the rate at which Rb^+ exited the lens. One would not expect to see an active component in Rb^+ efflux. Since they did not measure the PD between the inside of the lens and the bathing medium, one cannot rule out the possibility that the increased efflux of Rb^+ occurred because of a drop in the lens potential. Paterson (1970c), measuring Rb^+ efflux from the rabbit lens, also found a small increase in the Rb^+ efflux after the addition of ouabain. Though at the time this was an unexpected finding, subsequent work from Paterson's lab (1975) has shown that ouabain causes a rapid reduction in an electrogenic potential in rabbit lenses which would then reduce the passive forces tending to retain Rb^+. It is not clear at this time that this electrogenic pump argument can completely explain the observed effects of metabolic inhibitors on Rb^+ and K^+ efflux.

It would appear that the majority of the K^+ in the lens is exchangeable with K^+ in the bathing solution. Thoft and Kinoshita (1965b) showed in the rabbit lens that equilibration of ^{42}K was not complete even after 24 hours of incubation. They reported, however, that the ^{42}K level in the lens was still increasing at the end of the 24 hours. A similar finding was noted by Duncan (1969c) in the amphibian lens, where he reported that even though the exchange was not complete after 24 hours, the ^{42}K level in the lens was asymptotically approaching 100% of that in the bathing solution. Duncan (1969c) found that, for the most part, the K^+ efflux rate increased when the lens was depolarized by various ionic substitutions, as one would predict for a passive flux. He found, however, that the K^+ efflux increased more than predicted by the voltage change when he depolarized the lens by increasing the K^+ concentration in the bath. He hypothesized that a K^+ exchange diffusion mechanism might exist. He also noted that the K^+ efflux increased when the lens was exposed to Na^+-free solutions. From the hyperpolarization of the lens PD noted, one would have anticipated a decrease in efflux. At this point, the explanation for this phenomenon is not clear.

To date, there have been no satisfactory measurements of the passive efflux of K^+ from the separate anterior and posterior surface and thus no clear measurement of the relative passive K^+ permeabilities of the two sides of the lens. Candia *et al.* (1970, 1971) showed that the addition of K^+ to the bathing medium on the anterior side of the lens placed in an Ussing chamber caused a large change in the potential difference across that surface. Increasing the K^+ concentration in the posterior bath had little effect on the posterior potential. These results indicate that the anterior surface of the lens has a higher permeability to K^+ than the posterior surface. It is not possible at this point to quantitate the difference in these permeabilities. The data of Kinsey and Reddy (1965) from the rabbit do not support a difference between anterior and posterior passive K^+ permeability. They showed that in lenses incubated in 10^{-5} M ouabain, the ^{42}K uptake across both surfaces was equal. At this inhibitor concentration, active transport should have been so reduced that any ^{42}K movement would be passive. Thus, if the anterior surface K^+ permeability were greater than that for the posterior, greater accumulation should have occurred through the anterior. It did not. As previously stated, Kinsey and Reddy (1965) had found that when they separated the anterior and posterior surfaces and looked at the relative uptakes across the two surfaces, approximately three times as much K^+ moved across the anterior surface as the posterior in a similar incubation time. Since both surfaces appeared to have equal passive permeabilities, one would have to hypothesize the occurrence of a net flux of K^+ from anterior to posterior. Lack of a net flux could only occur if the posterior K^+ permeability were small in comparison to the anterior K^+ permeability, so as to offset the effect of the anterior pump. The predicted net flux of K^+ is also at odds with the results of Candia *et al.* (1970, 1971) where they found equal unidirectional fluxes in the Ussing chamber-mounted toad lens.

Although many of the criteria for active transport have been satisfied by the K^+ handling mechanisms in the lens, it is difficult to rule out the possiblity that the K^+ movement is, in fact, not active, but only follows as a result of the active extrusion of sodium ion as suggested by Candia (1973). Such a controversy has existed in other tissues for some time. One is tempted to call the obviously mediated K^+ movement *active transport* because of the involvement of the Na^+–K^+-activated ATPase which is inhibited by ouabain, but at the present time, no rigorous proof for the active transport of K^+ exists. Candia was unable to show a short circuit current due to K^+ ion in the toad lens. The unidirectional fluxes in that preparation were equal in both directions. Bito and Davson (1964) presented *in vivo* evidence that a net K^+ flux might occur in mammals. The failure to find a net K^+ flux across the toad lens does not rule out the possibility of active transport. Since it appears that the passive permeability of the posterior surface to K^+ is less than that of the anterior in the toad lens, the K^+ that might be pumped by an anterior epithelial pump could simply leak back through the anterior surface and result in no net translens flux. Therefore, the active transport of K^+ in the lens remains an unanswered question.

b. Sodium

Sodium fluxes into and out of the lens have been investigated by a number of individuals (Kinsey and Reddy, 1965; Duncan, 1969a, 1970b; Paterson, 1969b, 1970b,c; Kinsey, 1973a; Candia *et al.*, 1970, 1971). Kinsey and co-workers have also characterized the fluxes for other alkali metal cations (Kinsey *et al.*, 1971; Kinsey, 1973b; Kinsey and McLean, 1974). One would anticipate that the Na^+ influx would be entirely passive, whereas its efflux would be both active and passive. The importance of such a Na^+ transport system has been elegantly covered by Duncan and Croghan (1969). Paterson (1970b) and Duncan (1970b) have both shown that only 85% of the lens Na^+ is exchangeable with Na^+ in the bathing medium, a result previously reported by Harris and Gruber (1962) and Kinsey *et al.* (1958). Paterson showed that the efflux of ^{22}Na from the rabbit lens was remarkably inhibited by 10^{-5} *M* ouabain and lower temperature. He also showed that manual decapsulation of the lens rendered the ^{22}Na efflux insensitive to ouabain. These findings are in agreement with the concept that Na^+ is actively transported from the inside of the lens into the bathing medium across the lens epithelium. Upon examining the efflux of ^{22}Na from lenses preloaded with the isotope, both Paterson (1970c) and Duncan (1970b) found that the efflux had three components; one a very fast component presumably due to efflux from the extracellular space and capsule and two longer components whose origins are unclear. Earlier, Paterson (1970b) had shown that the inexchangeable fraction for Na^+ was much higher in the nucleus than in the cortex. A similar result was found by Duncan (1970b) in the amphibian lens when by ^{22}Na autoradiographic techniques, he showed the nuclear Na^+ to be more slowly exchangeable than elsewhere. Duncan, therefore, hypothesized that the two long components in the Na^+ efflux might be explained by a membrane barrier as well as diffusion of Na^+ in the bulk of the lens being limited by its exchange with complexed ions. The functional significance of this Na^+ binding in the lens is unknown. Presently available lens data would suggest that neither K^+ (Paterson *et al.*, 1974) nor H_2O (Neville *et al.*, 1974) is bound.

It would appear that the anterior and posterior surfaces of the lens have different permeabilities to Na^+. Kinsey and Reddy (1965) reported that when the two sides of the lens were isolated and ^{22}Na added to the bathing solution on only one side or the other, the influx of Na^+ across the posterior side was approximately four times greater than the influx across the anterior. A similar result was found by Paterson (1969) and recently by Yuge *et al.* (1973a,b,c). It is not clear whether the epithelium on the anterior side represents a real barrier or if the difference in this data might in part be due to the tortuous interdigitation between epithelial cells with a resulting increased diffusion path length. The tightness or barrier function of the epithelium is an area in which further research is needed. Candia *et al.* (1970, 1971) clearly showed that Na^+ was pumped from the inside of the lens across the anterior layers into the bathing solution by an active process. They showed that there was a short-circuit current for Na^+ from posterior to anterior. Na^+ was the only ion for which a short-circuit current and net flux could be demonstrated. No Na^+ efflux studies from a preloaded Ussing chamber-mounted lens have been reported in the literature. Such experiments might help to determine if there is an active Na^+ extrusion from the posterior surface, a process for which there is little supportive data at the present time.

c. *Chloride*

Kinsey and Reddy (1965) showed that the uptake of Cl^- was the same across both surfaces of the lens, an uptake that could not be inhibited by ouabain or iodoacetate. Paterson and Eck (1971) reported that some of the Cl^- in the lens was inexchangeable, with the percentage exchange being 99.3, 91.3, and 82.9 for the capsule, cortex, and nucleus, respectively. Duncan (1970b), however, found that Cl^- was 100% exchangeable in the toad lens. His studies suggested that Cl^- efflux was entirely passive, depending only on its electrochemical gradient. His finding that the efflux decreased substantially in Cl^--free bathing solutions led him to hypothesize an appreciable exchange diffusion for Cl^- in the amphibian lens. Candia *et al.* (1970, 1971) reported that no short-circuit current or net flux for Cl^- could be demonstrated across the Ussing chamber-isolated toad lens. Kinsey and Hightower (1976), using flux and PD measurements combined, showed Cl^- to be passively distributed in the rabbit lens. Therefore, at this time, most data are compatible with the notion that there is no active transport system for Cl^- and that it is passively distributed according to the electrical gradient between the inside of the lens and the outside.

d. *Summary*

In summary, it appears that there is an active process at the anterior surface of the lens which results in the extrusion of Na^+ and the uptake of K^+ and Rb^+. There seems to be no question that active transport of Na^+ ion occurs. However, even though the movement of K^+ and Rb^+ satisfies most of the criteria for active transport, there has been no direct demonstration of active transport of K^+. One cannot unequivocally dispute the argument that it is the Na^+ that is pumped and the K^+ is simply dragged in by the change in electrical forces. The data supporting K^+ pumping would, however, seem to be about as good as that for most other tissues. Recently, substantial data have been accumulated that point to a role of SH groups in both the active pumping and passive leak of lens ions (Epstein and Kinoshita, 1970a,b; Epstein, 1971; Kinoshita and Merola, 1973; Fukui *et al.*, 1973, 1976; Fukui, 1976; Giblin *et al.*, 1976). The data dealing with the possible extent of coupling between the movements of Na^+ and K^+ ion have been contradictory (Duncan and Croghan, 1969; Duncan, 1970b; Kinsey, 1973a). Existing data would seem to indicate that the anterior surface of the toad lens is more permeable to K^+ than is its posterior surface, but this result cannot be verified in the rabbit. The permeability of the posterior surface of the rabbit lens to Na^+ is greater than the corresponding anterior permeability. It would appear that the Cl^- in the lens is passively distributed and crosses either surface of the lens equally. It is known that all three, Na^+, K^+, and Cl^-, are asymmetrically distributed within the bulk of the lens. The extent to which this asymmetric distribution depends upon internal ion-limiting membranes, active transport processes, and ion binding phenomena is yet to be determined.

E. Origin of the Potential Difference

A biopotential can result either passively from a steady-state ionic distribution across a membrane or from an energy-requiring electrogenic pump. To say that a potential is passive does not mean that ion pumps are not involved somewhere in the process. It only means that the pump does not directly cause charge separation across the membrane. Such nonelectrogenic pumps are responsible for creating ionic imbalances across membranes, but the ions then assume a steady-state distribution.

Duncan (1969a,b,c; 1970a,b) and Duncan and Croghan (1969, 1970) published the results of a superb series of experiments in which they demonstrated that the PD of the toad lens was the result of a steady-state ionic distribution. Duncan measured influxes and effluxes of Na^+, K^+, and Cl^- and determined the effect of changes in the lens PD on these fluxes. Using the Goldman (1943) potential and flux equations or the more extensive comparable Kimizuka and Koketsu (1964) equations, he was able to show that in most instances, PD and conductance changes which occurred in the lens when the ionic content of its bath was altered were quantitatively very similar to those predicted by the theoretical equations for a passive distribution. The correspondence of data and theory was even more impressive when exchange diffusion was taken into account. His studies clearly established the necessity of measuring both lens PDs and ionic fluxes when attempting to understand the significance of ionic influx and efflux data.

Paterson *et al.* (1975) and Hightower and Kinsey (1977) were able to show that in the rabbit lens, the PD was too great to be predicted from steady-state ionic distributions. When Paterson *et al.* (1975) cooled the lens or treated it with $10^{-5} M$ ouabain, the PD quickly fell from −70 to about −52 mV, a value predicted by steady-state distributions. Upon rewarming, the lens recovered its initial PD. Ouabain was effective only when it was applied to the anterior surface. The authors then suggested that an electrogenic PD of 11–19 mV exists in the rabbit lens in addition to the 50–60 mV PD caused by steady-state ionic distributions. Hightower and Kinsey (1977) found similar results. In addition, by measuring Na^+ and K^+ diffusion rates, they ruled out an effect of temperature or ouabain on lens membrane permeability. Candia (1973) also reported an electrogenic Na^+ pump in the toad lens.

It is possible that the difference in the electrogenic pump findings between rabbit lenses and Duncan's toad lenses might simply be a species difference. It is also possible that in Duncan's studies, where he dissected lenses free of the globe, some trauma might have occurred. Rae (1973a), Paterson and Rae (1974), and Kinsey (1973a) have reported on PD and Na^+ permeability effects resulting from such a dissection procedure. Duncan's PD values of −63 mV for the toad lens is somewhat lower than the approximately −74 mV PD reported by Brindley (1956), Rae (1973a), Murata *et al.* (1974), and Paterson *et al.* (1974) for the frog lens. It would be of some interest to look for an electrogenic component in these higher PD frog lenses.

V. SUMMARY

Early investigations of the crystalline lens tacitly assumed it to behave like a giant red cell. Concentrations of materials inside were compared to their concentrations in the bathing solution without regard to possible asymmetric distribution within the lens. The "permeability" of the lens to various substances was determined also without regard to the actual location of the membranes responsible for the permeability. As a result of numerous studies over the past 15 years, we are now beginning to realize that the model of the lens as a giant, uniform, "spherical" cell is no longer adequate. Not only do ion asymmetries exist between the inside of the lens and its bath, but also asymmetries exist between different regions within the lens. Similar asymmetries exist for proteins and other metabolites. The active transport processes are clearly asymmetrically located. Even passive permeabilities seem asymmetrically located. It also seems highly probable that at least the superficial lens fibers are functioning cells that are surrounded by extracellular fluid. It is clear that cells within the lens are electrically coupled. Therefore, the spherical cell concept that the flux of all substances occurs only across the outer membranes of the outer-most layer of cells is highly questionable.

All of these observations place certain limitations on the lens experimenter and on the ease with which precise information can be extracted from the lens. When one measures influxes or effluxes of substances from cultured, totally immersed lenses, one obtains only a composite result which depends upon the individual contribution of each asymmetric pump and leak involved. If flux changes are noted, one cannot be sure which part of the lens changed. This means that one must begin thinking in terms of measuring fluxes in lenses where at least the anterior and posterior surfaces are isolated. Also, one cannot be sure of the total membrane surface area across which the fluxes are occurring. If the lens really has a functional extracellular space, fluxes may be occurring across the large total surface area of all fibers rather than across some much more limited area calculated from considering the lens as one giant cell. Since the lens has a relatively large PD between its interior and its bathing medium, changes in that PD must be taken into account when flux measurements are made. Both Duncan and Kinsey have made this point clearly in their publications.

The "interior" of the lens presents us with considerable questions. When substances penetrate into the lens do they penetrate through cells or between cells? Once in the extracellular space, is the resistance to their further movement in that space much lower than the resistance to movement through fiber membranes? Do substances entering one side of the lens get to the other side by diffusion in the extracellular space or by diffusion along the cytoplasm of fiber cells? How much movement is circumferential and how much, if any, occurs directly through the nucleus? Are there mechanisms for improving the rate at which substances move from one end of a fiber cell to the other through its cytoplasm? Are deep fibers leaky? Additional questions arise when one looks at the results of isotope efflux compartmental analyses. If ^{36}Cl or ^{42}K are used, two lens compartments are resolved. The estimate of the relative sizes of the two compartments is similar for these isotopes. If ^{22}Na or ^{24}Na is used, three compartments are found. If tritiated H_2O is used, again two compartments are found, but their relative sizes are very different than those found for ^{36}Cl or ^{42}K. What features of the lens interior could account for this curious compartmentalization?

These are but a few of the observations and questions which have arisen from various results presented in this review. It is clear that further insight into the function of the crystalline lens will result as investigators consider it to be a complex syncytium of coupled cells rather than one giant cell.

Acknowledgments

This work was supported by N.I.H Grants EY 01207 and NS 13778.

References

Amoore, J. E., Bartley, W., and Van Heyningen, R. (1959). *Biochem. J.* **72**, 126–133.
Andree, G. (1958a). *Pfluegers Arch.* **267**, 109–116.
Andree, G. (1958b). *Pfluegers Arch.* **267**, 117–119.

Aviram, A., Schalitt, M., Kassem, H., and Groen, J. J. (1966). *Clin. Chim. Acta* **14**, 442–449.
Bando, M., Nakajima, A., Nakagawa, M., and Hiraoka, T. (1976). *Exp. Eye Res.* **22**, 389–392.
Becker, B. (1962). *Invest. Ophthalmol.* **1**, 502–506.
Becker, B., and Cotlier, E. (1962). *Invest. Ophthalmol.* **1**, 642–645.
Becker, B., and Cotlier, E. (1965). *Invest. Ophthalmol.* **4**, 117–121.
Benedetti, E. L., Dunia, I., and Bloemendal, H. (1974). *Proc. Natl. Acad. Sci. U.S.A.* **71**, 5073–5077.
Benedetti, E. L., Dunia, I., Bentzel, C. J., Vermorken, A. J. M., Kibbelaar, M., and Bloemendal, H. (1976). *Biochim. Biophys. Acta* **457**, 353–384.
Bennett, M. V. L. (1973a). *Fed. Proc., Fed. Am. Soc. Exp. Biol.* **32**, 65–73.
Bennett, M. V. L. (1973b). *In* "Intracellular Staining in Neurobiology" (S. B. Kater and C. Nicholson, eds.), pp. 115–134. Springer-Verlag, Berlin and New York.
Bentley, P. J. (1973). *Exp. Eye Res.* **15**, 723–729.
Bentley, P. J., and Candia, O. A. (1971). *Invest. Ophthalmol.* **10**, 672–676.
Bentley, P. J., and Candia, O. A. (1975). *Am. J. Physiol.* **229**, 1520–1525.
Bentley, P. J., and Yorio, T. (1976). *Gen. Pharmacol.* **7**, 355–360.
Bettelheim, F. A. (1975). *Exp. Eye Res.* **21**, 231–234.
Bettelheim, F. A., and Vinciguerra, M. J. (1971). *Ann. N.Y. Acad. Sci.* **177**, 429–439.
Bettelheim, F. A., and Wang, T. J. Y. (1974). *Exp. Eye Res.* **18**, 351–356.
Bettelheim, F. A., Vinciguerra, M. J., and Kaplan, D. (1973). *Exp. Eye Res.* **15**, 149–155.
Bito, L., and Davson, H. (1964). *Exp. Eye Res.* **3**, 283–297.
Bloemendal, H., Zweers, A., Vermorken, F., Dunia, I., and Benedetti, E. L. (1972). *Cell Differ.* **1**, 91–106.
Bonting, S. L. (1965). *Invest. Ophthalmol.* **4**, 723–738.
Bonting, S. L., Caravaggio, L. L., and Hawkins, N. M. (1963). *Arch. Biochem. Biophys.* **101**, 47–55.
Brindley, G. S. (1956). *Brit. J. Ophthalmol.* **40**, 385–391.
Brown, D. A., Stumpf, W. E., and Roth, L. J. (1969). *J. Cell Sci.* **4**, 265–288.
Candia, O. A. (1973). *Exp. Eye Res.* **15**, 219–223.
Candia, O. A., Bentley, P. J., Mills, C. D., and Toyofuku (1970). *Nature (London)* **227**, 852–853.
Candia, O. A., Bentley, P. J., and Mills, C. D. (1971). *Am. J. Physiol.* **220**, 558–563.
Cohen, A. I. (1958). *Am. J. Anat.* **103**, 219–245.
Cohen, A. I. (1965). *Invest. Ophthalmol.* **4**, 806–814.
Cotlier, E. (1973). *Exp. Eye Res.* **15**, 711–722.
Cotlier, E., and Apple, D. (1973). *Exp. Eye Res.* **16**, 69–77.
Cotlier, E., Kwan, B., and Beaty, C. (1968). *Biochim. Biophys. Acta* **150**, 705–722.
Cotlier, E., Sanders, D., and Wyhinny, G. (1974). *Ophthalmic Res.* **6**, 107–130.
Cotlier, E., Baskin, M., and Kresca, L. (1975). *Invest. Ophthalmol.* **14**, 697–701.
Dainty, J., and House, C. R. (1966). *J. Physiol. (London)* **182**, 66–78.
Dickson, D. H., and Crock, G. W. (1972). *Invest. Ophthalmol.* **11**, 809–815.
Dickson, D. H., and Crock, G. W. (1974). *In* "Cataract and Abnormalities of the Lens" (J. C. Bellows, ed.), pp. 49–59. Grune & Stratton, New York.
Duncan, G. (1969a). *Exp. Eye Res.* **8**, 315–325.
Duncan, G. (1969b). *Exp. Eye Res.* **8**, 406–412.
Duncan, G. (1969c). *Exp. Eye Res.* **8**, 413–420.
Duncan, G. (1970a). *Exp. Eye Res.* **9**, 188–197.
Duncan, G. (1970b). *Exp. Eye Res.* **10**, 117–128.

Duncan, G. (1973). *In* "Ciba Symposium: The Human Lens in Relation to Cataract." Associated Scientif. Pubs., Amsterdam.
Duncan, G. (1974). *In* "The Eye" (H. Davson and L. T. Graham, Jr., eds.), 2nd ed., Vol. 5, pp. 357–397. Academic Press, New York.
Duncan, G., and Bushell, A. R. (1975). *Exp. Eye Res.* **20,** 223–230.
Duncan, G., and Bushell, A. R. (1976). *Exp. Eye Res.* **23,** 341–353.
Duncan, G., and Croghan, P. C. (1969). *Exp. Eye Res.* **8,** 421–428.
Duncan, G., and Croghan, P. C. (1970). *Exp. Eye Res.* **10,** 192–200.
Dunia, I., Sen Ghosh, C., Benedetti, E. L., Zweers, A., and Bloemendal, H. (1974). *FEBS Lett.* **45,** 139–144.
Eisenberg, R. S., and Johnson, E. A. (1970). *Prog. Biophys. Mol. Biol.* **20,** 1–65.
Eisenberg, R. S., and Rae, J. L. (1976). *J. Physiol.* **262,** 285–300.
Epstein, D. L. (1971). *Exp. Eye Res.* **11,** 351–355.
Epstein, D. L., and Kinoshita, J. H. (1970a). *Exp. Eye Res.* **10,** 228–236.
Epstein, D. L., and Kinoshita, J. H. (1970b). *Invest. Ophthalmol.* **9,** 629–638.
Farnsworth, P. N., Fu, S. C. J., Burke, P. A., and Bahia, I. (1974). *Invest. Ophthalmol.* **13,** 274–279.
Farnsworth, P. N., Burke-Gadomski, P., Kulyk, T., Mauriello, J. A., and Cinotti, A. A. (1976). *Exp. Eye Res.* **22,** 615–624.
Fahrquar, M. G., and Palade, G. E. (1963). *J. Cell Biol.* **17,** 375–412.
Fischer, F. P. (1933). *Arch. Augenheilkd.* **107,** 295–318.
Fowlks, W. L. (1965). *Invest. Ophthalmol.* **4,** 367–368.
Fowlks, W. L. (1973). *Experientia* **29,** 548–549.
Friend, D. S., and Gilula, N. B. (1972). *J. Cell Biol.* **53,** 758–776.
Frömter, E. (1972). *J. Membr. Biol.* **8,** 259–301.
Frömter, E., and Diamond, J. (1972). *Nature (London), New Biol.* **235,** 9–13.
Fukui, H. N. (1976). *Exp. Eye Res.* **23,** 595–599.
Fukui, H. N., Epstein, D. L., and Kinoshita, J. H. (1973). *Exp. Eye Res.* **15,** 249–253.
Fukui, H. N., Merola, L. O., and Kinoshita, J. H. (1976). *Doc. Ophthalmol. Proc. Ser.* **8,** 161–169.
Garcia, G. E., Aaberg, T. M., and Kupfer, C. (1964). *Invest. Ophthalmol.* **3,** 384–387.
Giblin, F. J., Chakrapani, B., and Reddy, V. N. (1976). *Invest. Ophthalmol.* **15,** 381–393.
Gilula, N. B. (1974). *In* "Cell Communication" (R. P. Cox, ed.), pp. 1–29. Wiley, New York.
Goldman, D. E. (1943). *J. Gen. Physiol.* **27,** 37–60.
Goodenough, D. A., and Revel, J. P. (1970). *J. Cell Biol.* **45,** 272–290.
Gorthy, W. C. (1968). *Exp. Eye Res.* **7,** 394–401.
Gorthy, W. C., Snavely, M. R., and Berrong, N. D. (1971). *Exp. Eye Res.* **12,** 112–119.
Hamai, Y., and Kuwabara, T. (1975). *Invest. Ophthalmol.* **7,** 517–527.
Hamai, Y., Fukui, H. N., and Kuwabara, T. (1974). *Exp. Eye Res.* **18,** 537–546.
Hansson, H. A. (1970). *Z. Zellforsch. Mikroskop. Anat.* **107,** 187–198.
Harding, C. V., Harding, D., Peters, V., Kuwabara, T., Reddan, J., Unakar, N., Bagchi, M., Schnur, T., and Gordon, S. (1973). *Exp. Eye Res.* **16,** 1–7.
Harding, C. V., Susan, S., and Murphy, H. (1976). *Ophthalmic Res.* **8,** 443–455.
Harris, J. E. (1966). *Trans. Am. Ophthalmol. Soc.* **64,** 675–699.
Harris, J. E., and Becker, B. (1965). *Invest. Ophthalmol.* **4,** 709–722.
Harris, J. E., Gehrsitz, L. B., and Nordquist, L. T. (1953). *Am. J. Ophthalmol.* **36,** 39–50.
Harris, J. E., and Gruber, L. (1962). *Exp. Eye Res.* **1,** 372–384.
Harris, J. E., and Nordquist, L. T. (1954). *Acta 17th Conf. Ophthalmol.* pp. 1002–1012.
Harris, J. E., Gehrsitz, L. B., and Nordquist, L. T. (1953). *Am. J. Ophthalmol.* **36,** 39–50.
Harris, J. E., Gruber, L., and Hoskinson, G. (1959). *Am. J. Ophthalmol.* **47,** 387–395.
Hart, W. M., Peckham, R. H., and Kimel, H. F. (1963). *Arch. Ophthalmol.* **69,** 76–82.

Heinrichs, D. J., and Harris, J. E. (1956). *Am. J. Ophthalmol.* **42**, 358–362.
Heinrichs, D. J., and Harris, J. E. (1957). *Arch. Ophthalmol.* **57**, 207–213.
Hightower, K. R., and Kinsey, V. E. (1977). *Exp. Eye Res.* **24**, 587–593.
Hogan, M. J., Alvarado, J. A., and Weddell, J. E. (1971). "Histology of the Human Eye." Saunders, Philadelphia, Pennsylvania.
Hollenberg, M. J., Wyse, J. P. H., and Lewis, B. J. (1976). *Cell Tissue Res.* **167**, 425–438.
Horowitz, P., Aronson, M., Grodzins, L., Ladd, W., Ryan, J., Merriam, G., and Lechene, C. (1976). *Science* **194**, 1162–1165.
Huggert, A. (1959a). *Acta Ophthalmol.* **37**, 522–527.
Huggert, A. (1959b). *Acta Ophthalmol.* **37**, 26–32.
Huggert, A., Jacobson, L., Odeblad, E., and Wikstrom, R. (1961). *Acta Ophthalmol.* **39**, 1048–1052.
Kaiser, R. J., and Maurice, D. M. (1964). *Exp. Eye Res.* **3**, 156–165.
Karnovsky, M. J. (1968). *In* "Biological Interfaces: Flows and Exchanges" (N.Y. Heart Assoc. Symp.) pp. 64–93. Little, Brown, Boston, Massachusetts.
Katz, B. (1966). "Nerve, Muscle, and Synapse." McGraw-Hill, New York.
Kern, H. L. (1965). *Invest. Ophthalmol.* **4**, 680–690.
Kern. H. L., Roosa, P., and Murray, S. (1962). *Exp. Eye Res.* **1**, 385–395.
Kimizuka, H., and Koketsu, K. (1964). *J. Theoret. Biol.* **6**, 290–305.
Kinoshita. J. H., and Merola, L. O. (1973). *Invest. Ophthalmol.* **12**, 544–547.
Kinoshita, J. H., Kern, H. L., and Merola, L. O. (1961). *Biochim. Biophys. Acta* **47**, 458–466.
Kinsey, V. E. (1965). *In* "The Structure of the Eye" (J. W. Rohen, ed.), Vol. II, pp. 383–394. Schattauer, Stuttgart.
Kinsey, V. E. (1973a). *Exp. Eye Res.* **15**, 699–710.
Kinsey, V. E. (1973b). *Invest. Ophthalmol.* **12**, 485–490.
Kinsey, V. E., and Frohman, C. E. (1951). *Arch. Ophthalmol.* **46**, 536–541.
Kinsey, V. E., and Hightower, K. R. (1976). *Exp. Eye Res.* **23**, 425–433.
Kinsey, V. E., and McGrady, A. V. (1971). *Invest. Ophthalmol.* **10**, 282–287.
Kinsey, V. E., and McLean, I. W. (1964). *Invest. Ophthalmol.* **3**, 585–591.
Kinsey, V. E., and McLean, I. W. (1970). *Invest. Ophthalmol.* **9**, 769–784.
Kinsey, V. E., and McLean, I. W. (1974). *Invest. Ophthalmol.* **13**, 784–794.
Kinsey, V. E., and Reddy, V. N. (1965). *Invest. Ophthalmol.* **4**, 104–116.
Kinsey, V. E., Wachtl, C., Kuck, J. F. R., and Reddy, V. N. (1958). *Acta 18th Conf. Ophthalmol.* pp. 865–872.
Kinsey, V. E., McLean, I. W., and Parker, J. (1971). *Invest. Ophthalmol.* **10**, 932–942.
Kuck, J. F. R. (1970). *In* "Biochemistry of the Eye" (C. N. Graymore, ed.), pp. 183–371. Academic Press, New York.
Kuwabara, T. (1968). *Arch. Ophthalmol.* **79**, 189–195.
Kuwabara, T. (1970). "Fine Structure of the Eye." Security-Columbian, Boston.
Kuwabara, T. (1975). *Exp. Eye Res.* **20**, 427–443.
Kuwabara, T., and Imaizumi, M. (1974). *Invest. Ophthalmol.* **13**, 973–981.
Kuwabara, T., Kinoshita, J. H., and Cogan, D. G. (1969). *Invest. Ophthalmol.* **8**, 133–149.
Lasser, A., and Balazs, E. A. (1972). *Exp. Eye Res.* **13**, 292–308.
Leeson, T. S. (1971). *Exp. Eye Res.* **11**, 78–82.
Ling, G. N., and Kromash, M. H. (1967). *J. Gen. Physiol.* **50**, 677–694.
Lowenstein, W. R. (1966). *Ann. N.Y. Acad. Sci.* **137**, 441–472.
McKay, G., Stewart, C. P., and Robertson, J. D. (1932). *Brit. J. Ophthalmol.* **16**, 193–201.
McLean, I. W., La Force, R. C., and Kinsey, V. E. (1969). *Doc. Ophthalmol.* **26**, 171–184.
McNutt, N. S., and Weinstein, R. S. (1970). *J. Cell Biol.* **47**, 666–688.
McNutt, N. S., and Weinstein, R. S. (1973). *Progr. Biophys. Mol. Biol.* **26**, 45–101.
Matsuto, T. (1973). *Acta Soc. Ophthalmol. Jpn.* **77**, 853–872.
Matuk, Y., Manery, J. F., and Dryden, E. E. (1969). *Can J. Physiol. Pharmacol.* **47**, 853–861.

Maurice, D. M., and Paterson, C. A. (1968). *J. Physiol. (London)* **196,** 53–54P.
Merola L. O., Kern, H. L., and Kinoshita, J. H. (1960) *Arch. Ophthalmol.* **63,** 830–835.
Moore, H. (1966). *Int. Rev. Exp. Pathol.* **5,** 179–216.
Murata, T., and Taura, Y. (1973). *Acta Soc. Ophthalmol. Jpn.* **77,** 877–880.
Murata, T., Taura, Y., and Akaike, N. (1974). *Exp. Eye Res.* **19,** 185–193.
Nelson, K. J., and Rafferty, N. S. (1976). *Exp. Eye Res.* **22,** 335–346.
Neville, M. C., Paterson, C. A., Rae, J. L., and Woessner, D. E. (1974). *Science* **184,** 1072–1074.
Nonaka, T., Nishiura, M., and Ohkuma, M. (1976). *J. Electron Microsc.* **25,** 35–36.
Palva, M., and Palkama, A. (1974a). *Acta Ophthalmol. Suppl.* **123,** 82–87.
Palva, M., and Palkama, A. (1974b). *Exp. Eye Res.* **19,** 117–123.
Palva, M., and Palkama, A. (1976). *Exp. Eye Res.* **22,** 229–236.
Pappas, G. D., and Purpura, D. P. (1972). "Structure and Functions of Synapses." Raven Press, New York.
Paterson, C. A. (1968). *Exp. Eye Res.* **7,** 431–433.
Paterson, C. A. (1969a). *Exp. Eye Res.* **8,** 442–446.
Paterson, C. A. (1969b). *Ann. Ophthalmol.* **1,** 346–348.
Paterson, C. A. (1970a). *Am. J. Physiol.* **218,** 797–802.
Paterson, C. A. (1970b). *Exp. Eye Res.* **10,** 151–155.
Paterson, C. A. (1970c). *Exp. Eye Res.* **10,** 331–338.
Paterson, C. A. (1972). *Doc. Ophthalmol.* **31,** 1–28.
Paterson, C. A. (1973). *Invest. Ophthalmol.* **12,** 861–863.
Paterson, C. A., and Eck, B. A. (1971). *Exp. Eye Res.* **11,** 207–213.
Paterson, C. A., and Maurice, D. M. (1971). *Am. J. Physiol.* **220,** 256–263.
Paterson, C. A., and Rae, J. L. (1974). *Exp. Eye Res.* **18,** 507–508.
Paterson, C. A., Neville, M. C., Jenkins, R. M., and Nordstrom, D. K. (1974). *Exp. Eye Res.* **19,** 43–48.
Paterson, C. A., Neville, M. C., Jenkins, R. M., and Cullen, J. P. (1975). *Biochim. Biophys. Acta* **375,** 309–316.
Pauly, H., and Schwan, H. P. (1964). *IEEE Trans. Biomed. Eng.* **11,** 103–109.
Peskoff, A., and Eisenberg, R. S. (1975). *J. Math. Biol.* **2,** 277–300.
Peskoff, A., and Ramirez, D. M. (1975). *J. Math. Biol.* **2,** 301–316.
Phillipson, B. (1969a). *Acta Ophthalmol. Suppl.* **103,** 3–30.
Phillipson, B. (1969b). *Invest. Ophthalmol.* **8,** 258–270.
Phillipson, B., Hanninen, L., and Balazs, E. A. (1975). *Exp. Eye Res.* **21,** 205–219.
Pirie, A. (1951). *Biochem. J.* **48,** 368–371.
Pirie, A., and Van Heyningen, R. (1956). "Biochemistry of the Eye." Blackwell, Oxford.
Rae, J. L. (1973a). *Exp. Eye Res.* **15,** 485–494.
Rae, J. L. (1973b). *Exp. Eye Res.* **15,** 697–698.
Rae, J. L. (1974a). *Invest. Ophthalmol.* **13,** 147–150.
Rae, J. L. (1974b). *Exp. Eye Res.* **19,** 227–234.
Rae, J. L. (1974c). *Exp. Eye Res.* **19,** 235–242.
Rae, J. L., and Blankenship, J. E. (1973). *Exp. Eye Res.* **15,** 209–217.
Rae, J. L., and Germer, H. A. (1974). *J. Appl. Physiol.* **37,** 464–467.
Rae, J. L., and Stacey, T. (1976). *Doc. Ophthalmol. Proc. Ser.* **8,** 233–239.
Rae, J. L., Hoffert, J. R., and Fromm, P. O. (1970). *Exp. Eye Res.* **10,** 93–101.
Rafferty, N. S. (1972). *Anat. Rec.* **173,** 225–228.
Rafferty, N. S., and Esson, E. A. (1974). *J. Ultrastruct. Res.* **46,** 239–253.
Rafferty, N. S., and Goosen, W. (1975). *Am. J. Anat.* **142,** 177–199.
Rafferty, N. S., Goosens, W., and March, W. F. (1974). *Am. J. Ophthalmol.* **78,** 985–995.
Reddy, V. N. (1972). *Exp. Eye Res.* **14,** 293–296.
Reddy, V. N. (1973). *Exp. Eye Res.* **15,** 731–750.

Revel, J. P., and Karnovsky, M. J. (1967). *J. Cell Biol.* **33**, C7–C12.
Riley, M. V. (1970). *Exp. Eye Res.* **9**, 28–37.
Sakuragawa, M., Kuwabara, T., Kinoshita, J. H., and Fukui, H. N. (1975). *Exp. Eye Res.* **21**, 381–394.
Salit, P. W. (1943). *Arch. Ophthalmol.* **30**, 255–258.
Sperelakis, N., and Potts, A. M. (1959). *Am. J. Ophthalmol.* **47**, 395–409.
Thoft, R. A., and Kinoshita, J. H. (1965a). *Exp. Eye Res.* **4**, 287–292.
Thoft, R. A., and Kinoshita, J. H. (1965b). *Invest. Ophthalmol.* **4**, 800–805.
Thoft, R. A., and Kinoshita, J. H. (1965c). *Invest. Ophthalmol.* **4**, 122–128.
Toyofuku, H., and Bentley, P. J. (1970). *Invest. Ophthalmol.* **9**, 959–965.
Ussing, H., and Zerahn, K. (1951). *Acta Physiol. Scand.* **23**, 110–127.
Van Heyningen, R. (1969). *In* "The Eye" (H. Davson, ed.), 2nd ed., Vol. 1, pp. 381–488. Academic Press, New York.
Vinciguerra, M. J., and Bettelheim, F. A. (1971). *Exp. Eye Res.* **11**, 214–219.
Von Sallmann, L., and Locke, B. D. (1951). *Arch. Ophthalmol.* **45**, 431–444.
Wakely, J. (1974). *Exp. Eye Res.* **18**, 571–577.
Wanko, T., and Gavin, M. A. (1958). *Arch. Ophthalmol.* **60**, 868–879.
Wanko, T., and Gavin, M. A. (1959). *J. Biophys. Biochem. Cytol.* **6**, 97–102.
Wanko, T., and Gavin, M. A. (1961). *In* "The Structure of the Eye" (G. K. Smelser, ed.), pp. 221–233. Academic Press, New York.
Weinstein, R. S., and McNutt, N. S. (1972). *N. Engl. J. Med.* **286**, 521–524.
Yamasaki, H. (1970). *Arch. Histol. Jpn.* **32**, 41–50.
Yorio, T., and Bentley, P. J. (1976). *Exp. Eye Res.* **22**, 195–208.
Yorio, T., and Bentley, P. J. (1976). *Exp. Eye Res.* **23**, 601–608.
Young, R. W., and Ocumpaugh, D. E. (1966). *Invest. Ophthalmol.* **5**, 583–593.
Yuge, T., Takeda, H., Imamura, A., and Tani, M. (1973a). *Exp. Eye Res.* **16**, 223–226.
Yuge, T., Takeda, H., and Tani, M. (1973b). *Acta Soc. Ophthalmol. Jpn.* **77**, 62–65.
Yuge, T., Takeda, H., and Tani, M. (1973c). *Acta Soc. Ophthalmol. Jpn.* **77**, 1987–1990.

CURRENT TOPICS IN EYE RESEARCH, VOL. 1

Structure and Function of Extraocular Muscle Fibers

D. J. CHIARANDINI[1]
J. DAVIDOWITZ[1]

I. Structural Characteristics 91
A. Introduction 91
B. EOM Fiber Populations 92
C. Innervation 99
D. Problems of EOM Sampling 101
E. Classification of EOM Fibers 102
II. Physiological Properties 107
A. Introduction 107
B. Electrical Properties of EOM Fibers 108
C. Contractile Properties of EOM Fibers 125
References 140

I. STRUCTURAL CHARACTERISTICS

A. Introduction

The extraocular muscles (EOM) are unusual in that they contain a wide assortment of muscle fiber populations, the morphological characteristics of which encompass virtually the entire range of striated muscle fiber categories described in the vertebrate musculature.

In 1971, Peachey comprehensively reviewed this literature from its histologic inceptions through the first several electron-microscopic descriptions and suggested that EOM fiber populations might be classified into five or six types. Since then, several studies have introduced dis-

[1] Department of Ophthalmology, New York University Medical Center, New York, New York.

ISBN 0-12-153001-9

agreements as to the number and characteristics of such populations in different species, and some aspects of the literature have become quite confused. Peachey's classification, though with some basic modifications, nevertheless remains as a convenient framework for summarizing the literature.

Hess (1961) and Hess and Pilar (1963) first demonstrated two basic classes of muscle cells in EOM, i.e., singly innervated fibers (SIFs) and multiply innervated fibers (MIFs). Each of these classes was later subdivided according to various parameters of internal cell structure and in terms of topographic localization within the two main cellular layers of oculorotatory EOM. In such a layer organization, first described by Kato (1938), a sheath of smaller diameter fibers covers the muscle's peripheral surface facing the orbital bone. This orbital surface layer (OSL) overlies the main body of the muscle, often referred to as the global region or layer. This second layer is composed of an admixture of small to large diameter fibers. MIFs and SIFs occur in both of these major muscle regions.

There is reasonably good agreement as to the major fiber types of the global region. Significant differences appear, however, among descriptions of the OSL in various species, particularly as regards the orbital MIFs. The endeavor of fiber classification recently acquired a further dimension of complexity with the demonstration that systematic morphological variability may occur along the length of individual muscle cells in some fiber populations, most notably so in orbital MIFs (Davidowitz *et al.*, 1977a,b). Insofar as prior fiber type definitions had been based on the tacit assumption of structural homogeneity throughout the muscle cell, the interpretation of reported species differences must now be tempered with uncertainties as to whether or to what extent such apparent differences might not be reflections of inadvertent sampling bias.

B. EOM Fiber Populations

The following outline will be limited to those investigations that attempted a comprehensive description of all the fiber populations within the muscle under study, and that included, among their defining variables, both innervational characteristics (single vs. multiple) and localization within the cellular layering.

The first of such studies was by Mayr (1971) on rat EOM. [An earlier description of monkey EOM by Miller (1967) did not specify fiber innervation, though it was suggested by Peachey (1971) that Miller's fiber categories are largely compatible with those of Mayr.] Subsequent studies include those in sheep superior rectus (Harker, 1972), cat superior rectus (Peachey *et al.*, 1974), cat inferior oblique (Alvarado and Van Horn, 1975), mouse superior rectus (Pachter *et al.*, 1976), and rabbit superior rectus (Davidowitz *et al.*, 1976a, 1977a,b).

Of the numerous cellular characteristics which have been employed for the definition of EOM cells, this outline will limit itself to those parameters which are common to all the studies to be considered, thus permitting generalized comparisons. Among the most evident ultrastructural characteristics in the transverse plane of section are (1) mitochondrial size and number, (2) fibrillar size and delineation, and (3) development of the sarcoplasmic reticulum (SR). Insofar as these basic structural features in large measure determine the overall cellular appearance we refer below solely to those three features. In some studies, emphasis is also placed on parameters of the longitudinal plane of section (M and Z lines) and/or on the histochemical profile. Such parameters are not critical, however, in the present context of broadly differentiating the respective fiber populations, and inclusion of such further details would only add unneeded complication to this necessarily concise review. For simplicity, additional details of muscle layering divisions described by some authors will be duly noted under the two major headings of global region and orbital surface layer. Only the extrafusal fibers of oculorotatory muscles will be considered.

1. Global Region

In all descriptions of the global region, the SIFs range from mitochondria-rich cells of typically small diameter to cells which are mitochondria-poor and of large diameter; these would be respectively analogous to the so-called red and white fibers of skeletal muscle (see Close, 1972). SIFs intermediate in these characteristics are also described in all studies except that of mouse (Pachter *et al.*, 1976). In cat (Peachey *et al.*, 1974; Alvarado and Van Horn, 1975), mouse (Pachter *et al.*, 1976) and rabbit (J. Davidowitz, unpublished observations), there is a tendency for mitochondria-poor SIFs to have greater SR development than those that are mitochondria-rich. In rat (Mayr, 1971) and sheep (Harker, 1972) the SR is well developed in all the global SIFs. The fibrils tend to be of comparably small size in the SIFs of the global region, except in mouse (Pachter *et al.*, 1976) where the fibrils are relatively larger in the mitochondria-poor SIFs, and in rabbit where the latter fibers are seen with either small or relatively large fibrils (J. Davidowitz, unpublished observations).

All studies have described an MIF that is characteristic of the global region. Such fibers display small, sparse mitochondria, a poorly developed SR largely confined to the I band, and fibrils that are large or indistinctly delineated. These MIFs are of medium–large diameter except in mouse where they are relatively small (Pachter *et al.*, 1976). In cat (Peachey *et al.*, 1974; Alvarado and Van Horn, 1975) and rabbit (Davidowitz *et al.*,

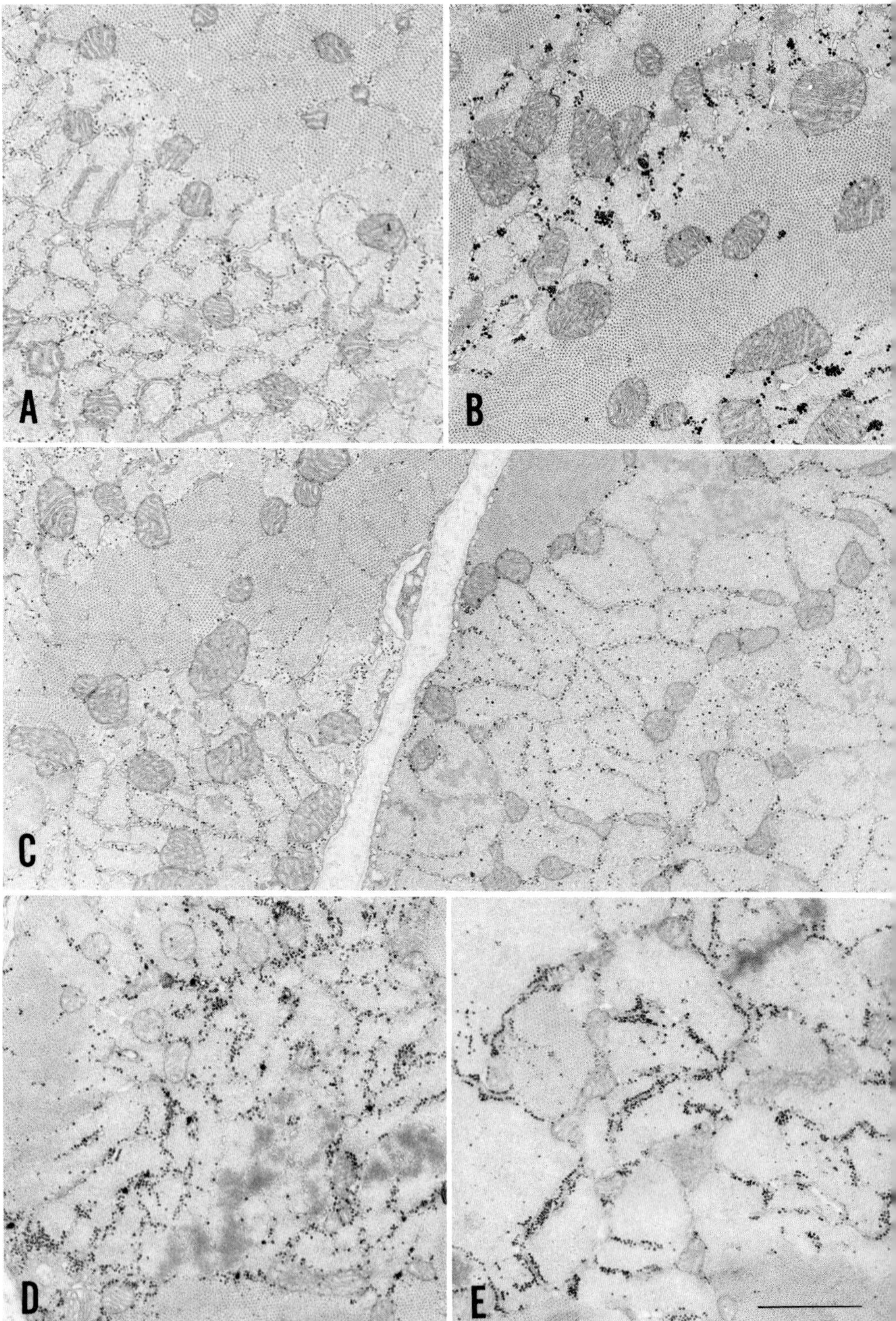
A
B
C
D
E

1977b) the global region also contains a second variety of MIF, namely that which is typical of the OSL. Figure 1 illustrates five varieties of global region fibers, three SIFs and two MIFs, as they appear in rabbit EOM (J. Davidowitz, unpublished observations).

2. Orbital Surface Layer

In comparing descriptions of the OSL, it will be helpful to consider as separate groups the first four studies (Mayr, 1971; Harker, 1972; Peachey *et al.*, 1974; Alvarado and Van Horn, 1975) and the last two studies (Pachter *et al.*, 1976; Davidowitz *et al.*, 1977a,b), insofar as procedural differences between these two groups may in part account for disparities in the respective findings. In the first group, the OSL was examined at several locations, but this standard sampling method makes no provision for identifying a given fiber at various portions of its length. In the second group, the OSL was serially sectioned thus permitting the sequential sampling of individual fibers.

a. Standard Sampling

All of the first four studies describe orbital SIFs containing numerous, large mitochondria. These cells have small fibrils that are delineated by

FIG. 1. Fiber populations in the global region of rabbit superior rectus. The transverse plane of section reveals differences in relative fibrillar size as well as in the content of sarcoplasmic reticulum (SR) and mitochondria. Parts (A), (B), and (C) illustrate four muscle cell categories of the muscle's belly region comparable to those described by most authors as the major "global" populations. These four cell categories in rabbit were not studied by serial analysis. (A) "White" SIF. A well-developed SR clearly delineates the fibrils in the I band (lighter area at the bottom half) and to a large extent in the A band (darker region at the top) as well. The fibrillar size shown here is small, but in some of these cells the fibrils may be moderately large. The mitochondria tend to be small and sparse. (B) "Red" SIF. A moderately developed SR delineates small fibrils in the I band; in the A band the SR is much reduced or virtually absent, as in the cell shown here. The mitochondrial content is high in terms of both size and number. (C, left) "Intermediate" SIF. The content of SR and/or mitochondria of such cells may fall almost anywhere within the wide range encompassed by the extremes shown in (A) and (B). (C, right) Large fibril MIF. The SR is characteristically reduced in the I band where the fibrils are large and/or ill-defined. Typically, the SR is virtually absent in the A band. The mitochondria tend to be small and relatively sparse. (D) "Pseudo-orbital" MIF, middle portion. This second variety of "global" MIF has been shown to be of variable morphology by serial analysis. Throughout most of the muscle's middle third, where these cells are of small (10 μm) diameter, their SR is moderately developed and delineates small fibrils in the I band which in this picture occupies most of the field except for narrow margins of A band at the left and bottom. Mitochondrial content is moderate to low. (E) Pseudo-orbital MIF, distal portion. The same cell as shown in (D) but at its distal portion where it increases to moderate diameter and merges with the orbital layer as other global fibers terminate. The SR is markedly reduced in the I band which occupies most of the field in this figure. Fibrils are ill-defined and/or vary widely in shape and size. The SR is virtually absent from the A band in both the middle and distal portions. Mitochondrial content is very low, particularly in the A band. Bar represents 1 μm (J. Davidowitz, unpublished observations).

moderate (Mayr, 1971; Peachey *et al.*, 1974; Alvarado and Van Horn, 1975) to large (Harker, 1972) amounts of SR in the I band. In the A band the SR tends to be poorly developed (Mayr, 1971; Peachey *et al.*, 1974; Alvarado and Van Horn, 1975). The diameter of these SIFs is typically small; in rat (Mayr, 1971) they are thicker toward the middle of their length.

In sheep (Harker, 1972) and cat (Peachey *et al.*, 1974; Alvarado and Van Horn, 1975), these orbital SIFs are classified as being of the same type as the global, mitochondria-rich SIFs. In rat, however, Mayr (1971) stresses that these global SIFs have a higher SR content than do their orbital counterparts, and he thus classifies them as two distinct types.

An additional intermediate SIF is described in the OSL of sheep (Harker, 1972), and in an apparent intermediate layer underlying the OSL in cat superior rectus (Peachey *et al.*, 1974).

The orbital MIFs are generally of small diameter (10–15 μm). They may also be very small in rat EOM (Mayr, 1971), while in sheep (Harker, 1972) MIFs of intermediate diameter are also seen at both ends of the muscles in "peripheral patches" overlying the OSL. Descriptions differ much in terms of mitochondrial content. In cat (Peachey *et al.*, 1974; Alvarado and Van Horn, 1975), the mitochondria are large and numerous, being comparable to those in the orbital SIFs. In rat (Mayr, 1971) and sheep (Harker, 1972) the mitochondria are small; they are sparse in the former, but more numerous in the latter. The SR is relatively sparse in all cases and tends to be confined to the I band where the fibrils are poorly delineated (Harker, 1972; Alvarado and Van Horn, 1975) or are moderate to large in size (Mayr, 1971; Harker, 1972; Peachey *et al.*, 1974).

b. Sequential Sampling

This procedure is based on the serial thick-sectioning of whole muscles embedded in Epon; such sections can be viewed by light microscope but remain available for electron microscopy (Davidowitz *et al.*, 1976b,c). Thus, individual fibers may be followed end to end, the number and location of their neuromuscular junctions may be ascertained, and the ultrastructure of a given muscle cell can be examined anywhere along its length.

This approach of virtually continuous sampling was first used for analysis of the OSL in rabbit superior rectus to clarify an apparently excessive variability of cell morphology encountered along this cell layer that could not be accounted for in terms of a reasonably small number of fiber types. Such analysis revealed that marked systematic variations of cellular structure occurs along SIFs and two major MIF populations (Davidowitz *et al.*, 1977a,b). The SIFs are of about 10 μm diameter with a localized swelling to about 15 μm near their end plates. One MIF is of relatively constant 10 μm diameter; the other is about 5 μm along its middle portion and increases to about 15 μm toward both ends (5–15 μm MIFs).

Near the middle of the muscle's proximal half, these SIFs and the two MIFs display localized elevations of mitochondrial content, most pronounced in the SIFs. For most of the OSL's extent, the SIFs and MIFs are essentially comparable in terms of small fibrillar size and moderate SR development in the I band; the A band is virtually devoid of SR, except in some SIFs where the SR becomes moderately developed here in association with the above noted increase in mitochondrial content.

As the 5–15 μm MIFs increase in diameter distally, their mitochondrial content becomes much reduced, particularly so in the A band. Most 5–15 μm MIFs extend into the muscle's distal third, where these fibers display a marked reduction in SR content whereby the fibrils may become large and/or indistinct. Such fibrillar transformation is not seen in the few 5–15 μm MIFs which terminate toward the end of the middle third, nor is it seen in any of the 10 μm MIFs all of which terminate before the end of the middle third; thus the absence of marked distal reduction in SR content in 10 μm MIFs would not necessarily represent a critical difference between these fibers and the 5–15 μm MIFs. A further similarity between 10 μm MIFs and 5–15 μm MIFs is evidenced by their parallel increase in the frequency of occurrence of membrane-glycogen complexes (Davidowitz *et al.*, 1975a,b) in the limited region of the OSL encompassed between the middle of the muscle and the beginning of its distal third (Davidowitz *et al.*, 1977b). Structural variations along rabbit OSL fibers are shown in Fig. 2.

An additional MIF population of the distal OSL is composed of the terminal portions of global MIFs; these are the only fibers that were analyzed serially in the global region. Such MIFs are of about 10 μm diameter as they traverse the muscle's middle portion within the global region. In the course of general fiber attrition toward the muscle's distal insertion, these MIFs gradually approach and eventually become an integral component of the OSL where their diameter increases to 15–20 μm. Associated with their distal diameter increase, these "pseudo-orbital" MIFs display a decrease in SR development, whereby the fibrils become ill-defined or large in size (see Fig. 1, D and E). Thus, these global MIFs appear quite distinct from the large-fibril global MIFs and show some similarity to the orbital 5–15 μm MIFs (Davidowitz *et al.*, 1977b).

In mouse superior rectus, however, serial analysis reveals that the orbital MIFs are of essentially constant morphology; they display small sparse mitochondria, moderately developed SR and small fibrils in the I band; the SR is virtually absent in the A band (Pachter *et al.*, 1976). These fibers increase in diameter from about 3 μm in the proximal third to about 7

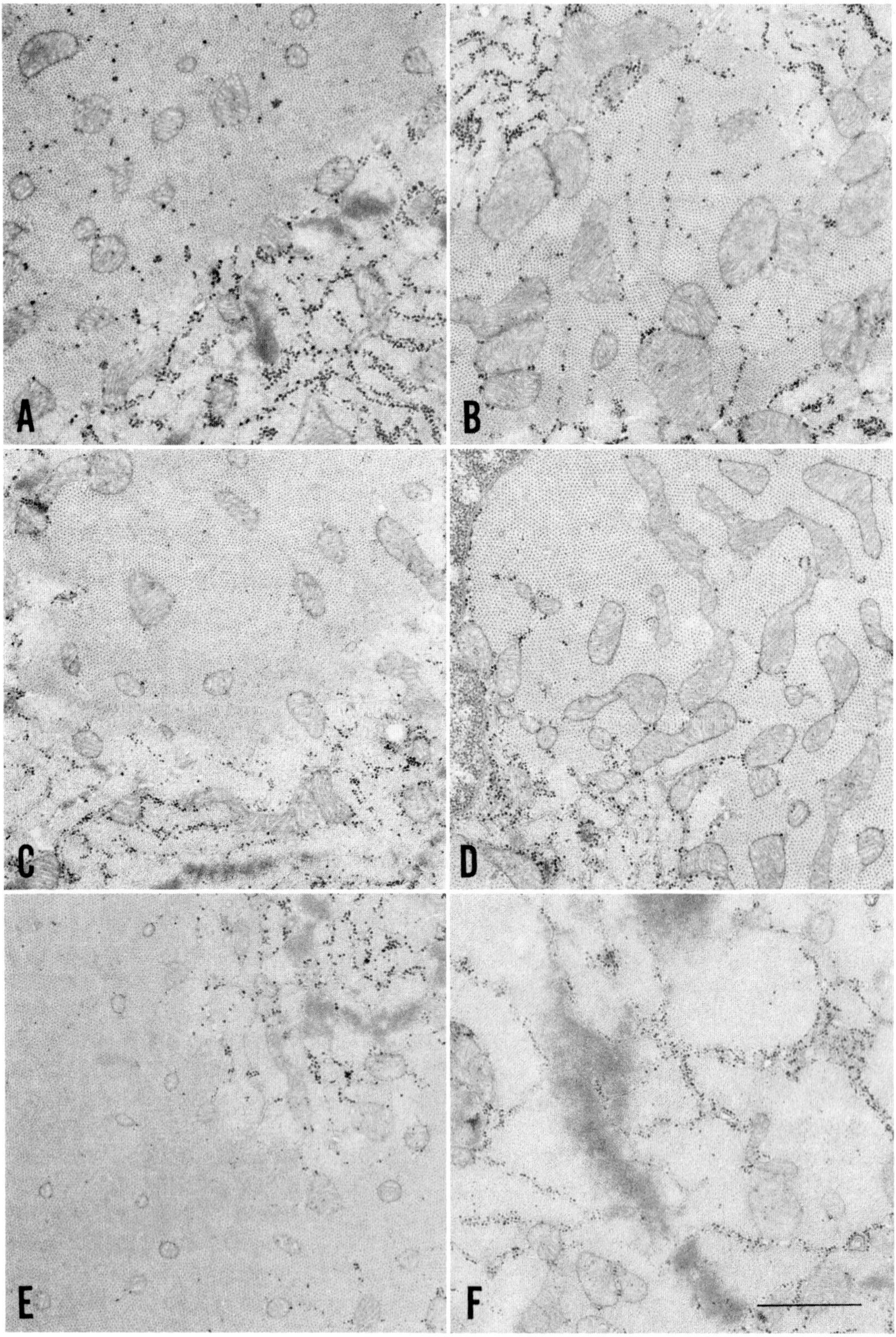
A
B
C
D
E
F

μm in the distal third. The orbital SIFs do vary in structure somewhat, with an increase in mitochondrial content in the middle third, where they are abundant and of medium size. Their fibrils are small and the SR is moderately developed in the I band; the A band is virtually devoid of SR except in the middle third where it may develop moderately. In their middle third, the orbital SIFs are about 10 μm in diameter and taper in either direction.

Serial analysis of the global region in mouse reveals no fibers with morphological variability (Pachter *et al.*, 1976). Thus, while structural variations occur along some MIFs of both the OSL and global region in rabbit, no such variations were seen in mouse.

C. Innervation

1. Types of Neuromuscular Junctions

Two basic types of neuromuscular junctions have been described in EOM (see Peachey, 1971, and Hess, 1974). In brief, the first type is the motor end plate, a junction wherein the terminal axon typically lies in a depression of the sarcolemma, which in turn may display varying degrees of postjunctional folding. There is general agreement that the SIFs are innervated by such motor end plates. In sheep, cat, and mouse EOM the large diameter mitochondria-poor SIFs of the global region are described as having more complex end plates in terms of number and/or depth of postjunctional foldings, as compared to the SIFs of smaller diameter and higher mitochondria content (Harker, 1972; Peachey *et al.*, 1974; Pachter

FIG. 2. Sequential transverse sections of SIFs (A, B) and 5–15 μm MIFs (C–F) of the orbital surface layer in rabbit superior rectus, as determined by serial analysis. (A) Portion of SIF well away from its end plate. Sarcoplasmic reticulum is moderately developed in the I band where it delineates small fibrils. Mitochondrial mass is moderate. (B) Portion of SIF near its end plate. The mitochondrial mass is markedly increased. In some of these fibers the SR may become moderately developed in the A band in this cell portion, as seen here. In most of the orbital SIFs, however, the SR remains virtually absent from the A band throughout the fiber length. (C) 5–15 μm MIF, as it appears along most of its middle segment of minimal diameter, which accounts for about a third of the fiber length. The SR is moderately developed in the I band where it delineates small fibrils. The SR is virtually absent from the A band of these fibers throughout their length. Mitochondrial content is moderate in A and I bands. (D) Portion of 5–15 μm MIF toward the proximal end of their minimal diameter segment, in the vicinity of an end platelike juction (sole plate nucleus is seen at left). Mitochondrial mass is much increased in A and I bands. Fibrils are small. (E) Portion of 5–15 μm MIF as it approaches its maximal diameter distally (toward end of muscle's middle third). Mitochondrial size and number are markedly reduced in the A band, and remain so till the distal termination. Fibrils tend to remain small here. (F) Portion of 5–15 μm MIF where diameter is maximal, in muscle's distal third. The SR is markedly reduced in the I band, where fibrils are ill-defined or vary widely in shape and size, often exceeding a width of 1 μm. Only the I band is shown here; the A band appears the same as in (E). This structural profile may account for as much as one-fourth of the fiber length. Bar represents 1 μm (Davidowitz *et al.*, 1977a,b).

et al., 1976). In the second type of junctions, the nerve ending lies superficially upon the sarcolemma and has either no or relatively minor postjunctional folding. Such essentially superficial neuromuscular junctions are generally thought to characterize the nerve terminations on multiply innervated fibers (Hess and Pilar, 1963; Mayr, 1971; Harker, 1972; Peachey *et al.*, 1974; Pachter *et al.*, 1976).

Teräväinen (1968) has described two types of multiple neuromuscular junctions in rat EOM by light microscopy, one being small and numerous while the other is large and relatively few in number; these two junctional types may also occur on the same fiber. By electron microscopy, he found both large and small neuromuscular junctions that are essentially superficial and display minimal postjunctional folding. Teräväinen (1968) indicated that some of the larger junctions may simply be close groupings of the smaller ones, but noted that his study did not warrant any definite decision as to whether the MIFs are also supplied by nerves ending in motor end plates, such as is the case in MIFs of muscle spindles (Barker, 1974).

Serial analysis of the OSL in rabbit superior rectus demonstrated that, in addition to multiple superficial junctions the 10 μm MIFs and 5–15 μm MIFs also display end platelike junctions; the latter resemble those seen in end plates of the SIFs in that the terminal axon lies in a scalloped depression of the sarcolemma that may develop numerous postjunctional folds (Davidowitz *et al.*, 1976a). These MIF end platelike junctions are apparently confined to that portion of the OSL which is encompassed by the SIF end plate band, i.e., toward the middle of the muscle's proximal half.

It is of interest that both SIF end plates and MIF end platelike junctions seem to be directly associated with the above mentioned localized increases of mitochondrial content that occur in the innervation band of the SIFs. Such increases might signify analogous cellular responses to these large types of nerve endings in both types of fibers. The electron microscopic observation of end platelike junctions in some rabbit MIFs brings to mind Hines' (1931) early light microscopic observation in rabbit EOM that a single fiber may carry both an end plate and a small accessory ending.

2. Motor Nerves

Studies of the innervation of rat EOM suggest that motor axons of MIFs are nonmyelinated, while those of SIF end plates are myelinated, implying a potential criterion for differentiating neuromuscular junctions on MIFs and SIFs, respectively (Teräväinen, 1968; Mayr, 1971). Serial analysis of the OSL in rabbit superior rectus indicated that this is not necessarily a reliable index for such differentiation since both myelinated and nonmyelinated nerves are seen in association with MIF neuromuscular junctions along the length of these fibers. Furthermore, some motor nerves may lose their myelin 10–15 μm before termination upon the fibers or while still within a nerve bundle (Davidowitz *et al.*, 1977b).

D. Problems of EOM Sampling

Before considering the question of classifying EOM fibers, it were well to temper our overview of the literature with an awareness of further complexities in these muscles which present potential pitfalls in their sampling. Peachey (1971) has stressed the significance of sample location along the muscle's length in view of the early finding by Hines (1931) that fibers may terminate in the middle third of rabbit EOM. Recent serial analysis of the OSL in rabbit, has confirmed Hines' observation for both SIFs and MIFs (Davidowitz *et al.*, 1977b); an anlogous situation has also been described in the OSL of mouse (Pachter *et al.*, 1976). Insofar as samples taken at various locations along the muscle might contain different fiber populations, the sample location would itself become a meaningful component of the definition of those populations. Thus, the importance of specifying sample location along the muscle would derive from its complex topographical organization, as well as from the above described (Section I,B,2) possibility that some SIFs and MIFs may display markedly differing morphologies as they travel from one portion of the muscle to another.

In EOM containing fibers displaying systematic morphological variability, however, it would seem that specification of where one is sampling, in terms of the muscle's overall length, may in itself be insufficient for fiber identification. In such fibers it would be also necessary to know the position of loci which apparently determine the variations of structure, and such loci may vary in position, within relatively wide limits, from one fiber to another, along the muscle's length. One apparently critical locus, described above (Section I,B,2), is the position of SIF end plates and MIF end platelike junctions which tend to be associated with localized increases in mitochondrial content. In the OSL of rabbit EOM (Davidowitz *et al.*, 1977b), such nerve endings are dispersed over as much as a fifth of the muscle's length (toward the middle of its proximal half). Thus a given plane of section within this rather broad innervation band, would contain fibers of the same population displaying respectively variant appearances which might be misinterpreted as representing different populations. Of particular significance for the MIFs would be their appearance near their end platelike junctions where the abundant mitochondria and small fibrillar size present a structural profile often ascribed to SIFs (see Section I,B).

A second variety of apparently critical locus, of particular significance in SIFs, would be their tapering ends; in the OSL of rabbit EOM, SIF tapers may occur almost anywhere in the proximal two-thirds of the muscle (Davidowitz *et al.*, 1977b). Within such tapers, there is often a marked decrease in the number and size of mitochondria in association with a reduction in SR content that results in enlarged and/or indistinct fibrils (J. Davidowitz, unpublished observations). Insofar as a comparable structural profile has often been ascribed to MIFs, this peculiarity of fiber origins and terminations may be a further source of misinterpretation in the attempt to identify EOM fiber populations. Indeed, the above varieties of apparently critical loci within OSL fibers may well combine in the belly region of the muscle, so that "SIF-like" portions of MIFs (near their end platelike junctions) may occur alongside "MIF-like" portions of SIFs (at their tapers), thereby opening the possibility of actually reversing their identification. Thus, an adequate characterization of component fiber populations, if not the avoidance of erroneous classification, would seem to require an awareness of sample location, both in terms of the muscle as a whole, as well as with respect to its position along each of the respective fibers.

E. Classification of EOM Fibers

In his 1971 review, Peachey suggested a system of classification based on five types, three SIFs and two MIFs, with an additional intermediate SIF as a tentative sixth category. These were based largely on the descriptions of EOM in monkey (Miller, 1967) and rat (Mayr, 1971). The data which has become available since that review has further complicated the problem of EOM fiber classification in two respects. First, divergencies among the respective descriptions raise the possibility of significant species differences and/or the presence of yet more fiber categories. Second, the occurrence of systematic morphological variability along the length of some MIFs and SIFs requires that such fibers be defined in terms of patterns of such change in their morphological profile, rather than by a set of more or less fixed structural parameters whereby EOM fiber types had been previously defined.

These added complexities notwithstanding, the fiber types enumerated by Peachey continue to be most serviceable as a point of departure for the classification of EOM fibers, and are thus retained as the bases for the following listing of six broad fiber categories. Though unenumerated as such by Peachey, the intermediate SIFs are here counted as a separate category, in view of their continued description in most studies. These third and fourth types; (his fourth and fifth types thus become the fifth and sixth categories of the present listing).

These six categories are shown in Table I, according to the terminology employed by the respective authors. Note that the same or analogous fiber populations may occur in both the OSL and global region, in some descriptions. Thus, whether or not the various categories of Table I represent mutually exclusive distinctive "fiber types" will vary according to the respective authors. The following comments associated with the six categories of Table I emphasize some discriminative aspects of the fuller descriptions presented above (Sections I, B, 1 and 2; pp. 93–99). Further details of clarification are given in footnotes of Table I.

1. Orbital SIFs

These are fibers of relatively small diameter and high mitochondrial content. According to two descriptions, however, the mitochondrial richness is preferentially concentrated toward the middle portions of the cellular length (Pachter *et al.*, 1976; Davidowitz *et al.*, 1977a,b), and three studies describe the diameters as being greater toward the middle of these cells (Mayr, 1971; Pachter *et al.*, 1976; Davidowitz *et al.*, 1977b). Insofar as such morphological and diameter variability has not been described in other SIFs, it is possible that they might differentiate this dominant population of the OSL from analogous fibers in the global region. It has been estimated that SIFs constitute about 70% of the OSL fibers (Mayr, 1971; Harker, 1972; Peachey *et al.*, 1974; Alvarado and Van Horn, 1975; Davidowitz *et al.*, 1977b). For simplicity, Table I omits the additional intermediate SIFs of the OSL described by Harker (1972).

2. Orbital MIFs

There are conspicuous differences among the respective descriptions of these fibers, particularly in terms of their mitochondrial and SR content. Insofar as serial analysis of orbital MIFs in rabbit superior rectus (Davidowitz *et al.*, 1977b) reveals marked systematic variation of mitochondrial content along these fibers, partly in relation to end platelike junctions, the possibility cannot be excluded that some of the discrepancies among authors might in part be related to sampling disparities. However, serial analysis of orbital MIFs in mouse superior rectus (Pachter *et al.*, 1976) finds neither morphological variability nor end platelike junctions in orbital MIFs, thus suggesting that real differences may exist in these fibers among the other species as well. The proportional representation of MIFs among OSL fibers is generally agreed to be about 20–35% (Mayr, 1971; Harker, 1972; Peachey *et al.*, 1974; Alvarado and Van Horn, 1975; Davidowitz *et al.*, 1977b). In Table I, a second orbital MIF population is also listed for sheep (Harker, 1972) and rabbit (Davidowitz *et al.*, 1977). Insofar as these authors do not classify these additional MIFs as specifically separate types to be differentiated from other MIFs of this layer, these second orbital MIFs are noted here, without separate numeration.

TABLE I

Fiber Populations in Extraocular Muscles[a]

	Mayr, 1971	Harker, 1972	Peachey *et al.*, 1974	Alvarado and Van Horn, 1974	Pachter *et al.*, 1976	Davidowitz *et al.*, 1977
	Rat EOM	Sheep superior rectus	Cat superior rectus	Cat inferior oblique	Mouse superior rectus	Rabbit superior rectus
Orbital surface						
1 SIF	Dark 10 →25 μm	Small C 15 μm	Small fibril granulated 15 μm	Type 3 12 μm	Orbital SIF 5→10 μm	Orbital SIF 10→15 μm
2 MIF	Clear 5–10 μm	Small G 13 μm	Large fibril granulated 15 μm	Type 5 12 μm	Orbital MIF 3→7 μm	10 μm MIF
		Intermediate G 24 μm				5→15 μm MIF
Global region						
3 SIF	Dark 10–20 μm	Small C 15 μm	Small fibril granulated 15 μm	Type 3 12 μm	Small dark 10–15 μm	Small dark 10–15 μm
4 SIF	Intermediate 20–30 μm	Intermediate C 25 μm	Lightly granulated 25 μm	Type 2 22 μm		Intermediate 15–25 μm
5 SIF	Pale 30–40 μm	Large A 29 μm	Fibrillar white 35 μm	Type 1 31 μm	Large light 20–25 μm	Large white 25–35 μm
6 MIF	Clear 15–25 μm	Large G 32 μm	Afibrillar white	Type 4 23 μm	Global MIF 5–7 μm	Large fibril MIF 15–20 μm
2 MIF			Large fibril granulated 15 μm	Type 5 12 μm		Pseudoorbital MIF 10→15 μm

[a] Diversities of nomenclature stem from their being based variously on light vs. electronmicroscopic appearance, histochemical profile, diameter pattern, topographical localization, or simply a sequential numerical listing. Further clarification of this nomenclature problem would be beyond the scope of this review.

Cell diameters designated as single numbers refer either to approximations, averages, or extremes of range, as variously described by respective authors. Diameters designated by two numbers, which are separated by a dash, refer to a characteristic range. Fibers which are described as varying in diameter, along their length, are so designated by two numbers which are separated by an arrow.

As noted in the text, this table omits an intermediate SIF located in the orbital surface layer of sheep superior rectus (Harker, 1972).

The "intermediate G" MIFs, described by Harker (1972), occur in "peripheral patches" which are superimposed on the proximal and distal portions of the orbital surface layer. They are here listed as orbital fibers, for simplicity.

Omitted from this table is reference to an apparent intermediate layer located between the orbital surface layer and global region of cat superior rectus (Peachey *et al.*, 1974). This intermediate layer contains both the orbital SIFs and MIFs as well as "lightly granulated" SIFs which presumably correspond to the intermediate SIFs.

This table and the text do not refer to short fibers which may be localized toward either end of the muscle. Relatively little information is available regarding such fibers, nor is it clear whether they might represent distinctive categories. The reader is referred, in this connection, to Mayr, 1971; Harker, 1972; Pachter *et al.*, 1976; Davidowitz *et al.*, 1977.

The respective studies differ regarding the question of whether the same "fiber type" appears in both the orbital surface layer and global region, as follows.

Mayr (1971): Each of the six categories is considered to be a distinct fiber type.

Harker (1972): The orbital "small C" SIF and "intermediate C" SIF (unlisted) are respectively of the same types as their global SIF counterparts. The orbital "small G" MIF and "intermediate" MIF (peripheral patches), though of somewhat differing morphology, are considered to be of essentially the same type, insofar as it was speculated that the former may derive from the latter by fiber splitting.

Peachey *et al.* (1974): The orbital "small fibril granulated" SIF and "large fibril granulated" MIF are respectively of the same types as their global counterparts; as previously noted, these two fiber types also appear in an unlisted apparent intermediate layer. The latter layer also contains a "lightly granulated", presumably intermediate SIF which is of the same type as its global counterpart.

Alvarado and Van Horn (1974): The orbital "Type 3" SIF and "Type 5" MIF are respectively of the same types as their global counterparts.

Pachter *et al.* (1976): Each of the 5 categories is considered to be a distinct fiber type.

Davidowitz *et al.* (1977): Though the orbital 10 μm MIFs and 5–15μm MIF differ in various respects, some similarities between these two fiber populations suggest that they may nevertheless be variants of the same basic type. The orbital "5–15 μm MIFs" and the global "pseudoorbital MIFs" also would appear to be variants of the same fiber type.

3. Global "Red" SIFs

These mitochondria-rich fibers of relatively small diameter are classified as being of the same type as orbital SIFs by some authors; others consider them to be distinctly different populations. While these global SIFs are of an apparently constant diameter in rat (Mayr, 1971) and mouse (Pachter *et al.*, 1976), thus differentiating them from the orbital SIFs of variable diameter in those species, it is nevertheless possible that this population of global SIFs does vary in diameter at different regions of the muscle in other species (see Peachey, 1971). We might note here again that in mouse superior rectus (Pachter *et al.*, 1976), the only muscle in which both the OSL and global region have been serially analyzed, these global SIFs are described as having a constant morphology along their length in contrast to the morphological variability along the orbital SIFs. These cells account for about 25–30% of the global SIFs (Mayr, 1971; Alvarado and Van Horn, 1975).

4. Global "Intermediate" SIFs

SIFs wherein several structural characteristics fall somewhere in between those of the preceding "red" and the following "white" SIFs apparently occur in the global region of most species. The notable exception is that of mouse (Pachter *et al.*, 1976), the smallest and possibly the least complexly organized among the EOM considered here. Mayr (1971) has raised the question of whether intermediate SIF may not represent but the central component of a more or less continuous spectrum of SIFs in the global region, a significant possibility that still requires further clarification (see Peachey, 1971). This intermediate category would tend to be a somewhat tenuous entity, if only because of the possibility for overlap with either "red" or "white" SIFs in terms of parameters such as fiber diameter, mitochondrial content, and SR development.

5. Global "White" SIFs

These are large diameter fibers with a relatively low mitochondrial content and a well-developed SR in both the I and A bands. Their fibrils are small in most instances, but may also be relatively large. But, the combination of large diameter, low mitochondrial content, and large fibrils may also occur in global MIFs (the following sixth category). Indeed, the latter fibers have been referred to as "white" fibers in one description (Peachey *et al.*, 1974). The highly developed SR and distinct fibrils of the white SIFs, however, clearly differentiate them from global MIFs which tend to have a poorly developed SR and rather indistinct fibrils. The proportion of these fibers among the global SIFs has been estimated to be between 30 and 40% by Mayr (1971), Harker (1972), and Alvarado and Van Horn (1975).

6. Global MIFs

All studies provide rather similar descriptions of a global MIF population characterized by large and/or indistinct fibrils, and a relatively low SR and mitochondrial content. The diameter of such fibers, however, may variously be among the smallest or largest seen in this region of the muscle. Fibers with this morphological profile, not unsimilar to that of the "slow" MIFs of frog (see Peachey, 1971), have been estimated to represent between 7 to 15% of the global fiber population (Mayr, 1971; Harker, 1972; Alvarado and Van Horn, 1975). As previously noted, however, these are not necessarily the only global MIFs, insofar as some authors also describe a second global MIF population that is of significantly different morphology; the latter are listed as class 2 at the bottom row of Table I, signifying that they are the same or analogous to MIFs seen in the OSL of these muscles. The proportion of these fibers has not been specified.

II. PHYSIOLOGICAL PROPERTIES

A. Introduction

Two pioneering studies published almost half a century ago showed that extraocular muscles have unique contractile properties. Cooper and Eccles (1930) demonstrated that these muscles were the most rapidly contracting of all mammalian muscles, having only a fraction of the contraction time of the fastest limb muscle that they studied. In contrast, Duke-Elder and Duke-Elder (1931) discovered that extraocular muscles developed a sustained or tonic tension when exposed to cholinomimetics. This slow contractile response is absent in other mammalian muscles but it is characteristic of denervated mammalian muscles and of most amphibian muscles. Thus, eye muscles appear to be highly specialized and to combine the physiological properties of different types of skeletal muscles.

The duality of the contractile properties of eye muscles was explained by the anatomical work of Hess (1961) who demonstrated in these muscles the presence of two main types of fibers that receive single or multiple innervation and have morphological characteristics comparable to those of twitch and tonic fibers of lower vertebrates. The morphological complexity of extraocular muscle fibers has been indicated in the previous section of this review. In this section the muscle cells will be discussed from a physiological point of view, analyzing initially their electrical characteristics and afterwards, their contractile properties. Previous reviews of these aspects of extraocular muscles can be found in Breinin (1971), Bach-y-Rita (1971) and in the symposium "Basic Mechanisms of Ocular Motility and Their Clinical Implications," edited by Lennerstrand and Bach-y-Rita (1975).

B. Electrical Properties of EOM Fibers

Numerous studies of the eye muscles of several species have shown the existence of two main types of fibers which can be classified electrophysiologically by the passive properties of the cell membrane, the presence or absence of action potentials, and the characteristics of the synaptic activity. The two types are (a) singly innervated with action potentials, or twitch fibers, and (b) multiply innervated without action potentials, or tonic fibers. The existence of a third type of fibers with multiple innervation and action potentials has been postulated by Bach-y-Rita and Ito (1966). Although the terms fast and slow fibers are frequently found in the literature to designate twitch and tonic fibers respectively, these terms are avoided here because they are also used to distinguish two subtypes of twitch fibers in mammalian muscles and thus lead to confusion. Lennerstrand (1974) adopted a nomenclature that is clear and conveniently brief. It is based on the type of innervation of the fiber and the presence or absence of action potentials. His classification corresponds to the one previously mentioned, as shown by the following: SI (singly innervated) fibers are the singly innervated or twitch fibers, MINC (multiply innervated nonconducting) fibers are the multiply innervated fibers without action potentials or tonic fibers, and MIC (multiply innervated conducting) fibers are the multiply innervated fibers with action potentials.

1. Twitch, Singly Innervated, or SI Fibers

Studies on the electrical properties of this fiber type have been carried out *in vivo* and *in vitro* in rabbit, cat, and rat (Hess and Pilar; Matyushkin, 1964; Bach-y-Rita and Ito, 1966; Lennerstrand, 1974; D. J. Chiarandini and E. Stefani, unpublished results). The resting potential of these fibers has been measured with intracellular microelectrodes and, as in other mammalian twitch fibers, ranges between −65 and −90 mV. Twitch fibers respond with action potentials to depolarizations induced by nerve stimulation or by electrical current applied intra- or extracellularly. The characteristics of the action potential are similar to those of the spike recorded in other twitch fibers. In cells with a normal resting potential (−65 to −90 mV) the action potential has a threshold of between −60 and −55 mV and an overshoot of about 20–30 mV. Sodium is the main ionic species involved in the depolarizing phase of the spike, as indicated by the reversible blocking action of tetrodotoxin (D. J. Chiarandini and E. Stefani, unpublished results).

The passive electrical properties of the twitch fibers of extraocular muscle have been examined *in vivo* in cat by Bach-y-Rita and Ito (1966) and *in vitro* at 20°–25°C in rat by D. J. Chiarandini and E. Stefani (unpublished results). The first authors impaled fibers of inferior oblique muscles with a single microelectrode connected to a bridge circuit, or with a double barreled microelectrode to simultaneously record from and inject polarizing current into the fibers. In rat inferior rectus, Chiarandini and Stefani measured membrane characteristics with two separate microelectrodes using the square pulse analysis described by Fatt and Katz (1951). With this technique erroneous measurements due to undetected imbalances of the bridge circuit could be avoided.

The values for the input resistance reported in the two mentioned studies are remarkably similar: 1.12 ± 0.30 MΩ (mean ± standard error) in 32 fibers of cat inferior oblique and 1.62 ± 0.22 MΩ in 22 fibers of rat inferior rectus. The time constant, however, was found to be different: 10.6 ± 3.62 msec in cat and 3.2 ± 0.4 msec in rat. Using the cable equations (Hodgkin and Rushton, 1946) and assuming the fiber diameter, the following values of the specific membrane resistance (R_m) were obtained: 12,500 $\Omega\text{-cm}^2$ in cat for a cell diameter between 30 and 45 μm (Bach-y-Rita and Ito, 1966) and 4,000 $\Omega\text{-cm}^2$ for a 30 μm fiber in rat (D. J. Chiarandini and E. Stefani, unpublished results). From the time constant and R_m, the values of the specific membrane capacity (C_m) were derived: 0.86 $\mu F/cm^2$ and 0.95 $\mu F/cm^2$ in cat and rat, respectively.

The values for the R_m and C_m in the extraocular muscles are respectively higher and lower than the values obtained in limb muscles of cat and rat. In comparable experiments R_m was found to be 1,430 $\Omega\text{-cm}^2$ and 540 $\Omega\text{-cm}^2$ and C_m, 3.5 $\mu F/cm^2$ and 4.8 $\mu F/cm^2$, respectively (Boyd and Martin, 1959; Kiyohara and Sato, 1967). These discrepancies may reflect genuine characteristics of the extraocular muscles or a misjudgment of the fiber diameter assumed for the calculations that could introduce a large error in the computation, since R_m is proportional to (diameter)^3. For the calculation of R_m of limb muscle fibers the diameter was not assumed because the space constant was known. There is no information about the resting ionic permeability of extraocular muscle fibers but indirect evidence suggests that singly innervated fibers of rat inferior rectus are permeable to both potassium and chloride (Chiarandini, 1976).

Two types of synaptic activities, spontaneous and evoked, are exhibited by singly innervated fibers of extraocular muscles. They are similar to the synaptic events of twitch fibers in other vertebrate skeletal muscles. High gain recording with intracellular microelectrodes placed in the vicinity of the end plate region reveals the presence of miniature end plate potentials (mepps), spontaneous and transient depolarizations of the cell with amplitudes of usually no more than 3 mV. The mepps are produced by the interaction of one synaptic vesicle or quantum of acetylcholine (ACh) with the cholinergic receptors located in the postsynaptic membrane.

The second kind of synaptic activity is the end plate potential (epp), a large and transient depolarization of the fiber that is produced by the simultaneous release of many quanta of ACh. This is brought about by the arrival at the nerve ending of an action potential originating in the motorneuron or evoked experimentally by nerve stimulation.

The characteristics of the mepps of twitch fibers have been studied in cat and rat extraocular muscles with intracellular recording (Hess and Pilar, 1963; Bach-y-Rita and Ito, 1966; D. J. Chiarandini and E. Stefani, unpublished results). Mepps can be recorded only in the area of the muscle where the nerve enters and branches. Their amplitudes have a normal distribution and their rise and half-decay times vary within a narrow range. The mepps occur at random and are blocked by D-tubocurarine. Figure 3 illustrates mepps recorded in the inferior rectus of rat *in vitro* and Table II summarizes the results obtained by different authors in extraocular and other muscles.

Under normal conditions the amplitude of the epp is such that during its rising phase it triggers an action potential which then masks the epp. Therefore, to study the characteristics of the epp it is necessary to depress the synaptic transmission until the epp is smaller and is no longer capable of triggering action potentials. This can be done by using curarelike agents, by lowering Ca, or increasing Mg in the extracellular fluid (Fatt and Katz, 1951; del Castillo and Katz, 1954), or by fatiguing the neuromuscular junction with repetitive stimulation (Pilar, 1967). In records obtained close to the end plate focus, Hess and Pilar (1963) have shown in cat superior oblique *in vitro* that the epp has a rise time between 2 and 3 msec and decays to half amplitude in 2–4 msec. Similar results have been obtained

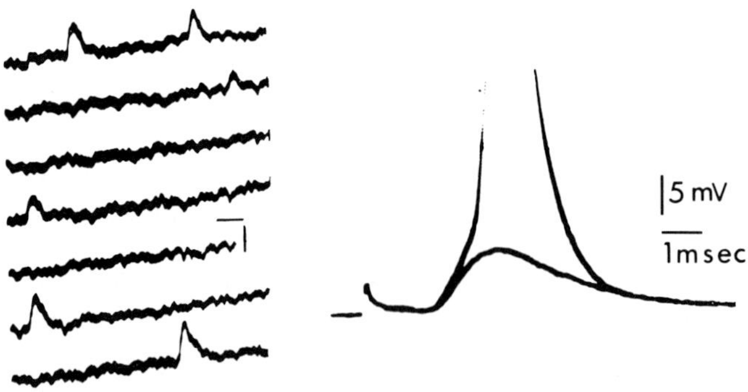

FIG. 3. Spontaneous and evoked synaptic activity in singly innervated fibers of rat inferior rectus muscle. Intracellular recording. (Left) Spontaneous miniature end-plate potentials have various amplitudes but rather similar rise times. Calibration: 1 mV, 10 msec. (Right) Muscle in which the synaptic transmission has been partially blocked with D-tubocurarine. Nerve stimulation evoked an action potential which went off the screen. A second shock applied immediately evoked only an end-plate potential (D. J. Chiarandini and E. Stefani, unpublished results).

TABLE II

Characteristics of Miniature End-Plate Potentials

Preparation	Voltage (mV)	Frequency (sec^{-1})	Rise time (msec)	Half-decay time (msec)	
Cat superior oblique *in vitro*, 20°–25°C	0.25–2.0	1.0–3.4	0.5–4.0	0.8–3.5	Hess and Pilar (1963)
Cat inferior oblique *in vivo*	0.41–1.37	5.0–31.0	0.3–0.9	1.93	Bach-y-Rita and Ito (1966)
Rat inferior rectus *in vitro*, 20°–25°C	0.3–2.5	1.2–11.0	0.3–4.5	1.0–4.4	D. J. Chiarandini and E. Stefani (unpublished results)
Cat tenuissimus *in vitro*, 35°C	0.3–0.9	0.5–5.0	0.4–1.2	0.8–2.0	Boyd and Martin (1956)
Rat diaphragm *in vivo* and *in vitro*, 37°C	0.5–6.0	1.0–6.0	0.8–2.0	1.3–1.9	Liley (1956)

in rat inferior rectus and are shown in Fig. 3 (D. J. Chiarandini and E. Stefani, unpublished results).

2. Tonic, Multiply Innervated Fibers without Action Potentials or MINC Fibers

The work of Hess and Pilar (1963) in cat extraocular muscles demonstrated the existence of fibers that are multiply innervated and have electrical and contractile properties different from those of singly innervated fibers. These authors showed that multiply innervated fibers did not generate action potentials, but upon repetitive electrical stimulation these fibers developed graded tension of a tonic or sustained nature. In many respects these fibers are comparable to the slow or tonic fibers or amphibian muscle described by Kuffler and Vaughan Williams (1953a,b). The existence of a similar type of multiply innervated fibers was demonstrated in rabbit extraocular muscles by Matyushkin (1964), who was also able to identify their motor response by selectively stimulating the small diameter axons that provide their innervation.

These studies of the extraocular muscles demonstrated for the first time that the multiply innervated fibers recognized at a morphological level by Kupfer (1960) and Hess (1961) had distinct physiological properties and explained the unusual contractile responses of these muscles to cholinergic drugs described by Duke-Elder and Duke-Elder (1931) and by Brown and Harvey (1941).

The values of resting potential in cat extraocular muscles reported by Hess and Pilar (1963) for tonic or MINC fibers were in general lower (less negative) than those of twitch fibers. A similar difference has been observed in rabbit (Matyushkin and Drabkina, 1970) and, in some cases, in rat (D. J. Chiarandini and E. Stefani, unpublished results). In amphibian muscles, multiply innervated fibers also have a relatively low resting potential. In these fibers Stefani and Steinbach (1969) have demonstrated that the low resting potential is due to the presence of a shunting resistance around the micropipette insertion and that with carefully selected microelectrodes it is possible to measure resting potentials comparable to those of the singly innervated fibers. The recording conditions approximate an equivalent circuit in which the shunting resistance (R_s) can be represented by a resistor in parallel to the input resistance of the fiber (R_{inp}) which is connected in series to the battery generating the resting potential of the cell (V). In such a circuit only a fraction of V (V_s) is recorded with the micropipette across R_s. The value of V_s is given by

$$V_s = \left(\frac{V}{R_{inp} + R_s} R_s \right)$$

In multiply innervated frog muscle fibers, R_s has been estimated to be about 20 MΩ, and since in these fibers the mean R_{inp} is 4.4 MΩ, a measured resting potential of −65 mV reflects a resting potential that, before the impalement, was about −80 mV (Stefani and Steinbach, 1969). From the equation it is clear that the attenuating influence of R_s on V is observed when R_{inp} is relatively large with respect to R_s.

As will be further discussed, multiply innervated fibers of extraocular muscles also have a large R_{inp}, so it is predictable that the shunting resistance around the microelectrode will have a marked attenuating effect on their resting potential. However, it is not possible to regard a low value of resting potential as a distinguishing characteristic since this parameter could vary with the magnitude of micropipette shunting which should depend on the size and shape of the micropipette tip, the angle of penetration of the micropipette, etc.

The electrical characteristics of the membrane of tonic of MINC fibers have been explored in a very limited way, but it is clear that they differ sharply from those of the twitch or singly innervated fibers and are quite similar to those of the multiply innervated fibers of amphibian muscle. Hess and Pilar (1963) measured with a bridge circuit the input resistance of some tonic fibers in cat superior oblique and reported a value of 7 MΩ. In rat inferior rectus, D. J. Chiarandini and E. Stefani (unpublished results) impaled with two microelectrodes several tonic fibers and obtained a mean input resistance of 6.0 ± 1.1 MΩ (mean ± standard error). These high values of input resistance contrast with the lower values mentioned previously that have been reported for singly innervated fibers of extraocular muscles. The high values of input resistance of these multiply innervated fibers could be explained by a very low membrane permeability, to Cl, a possibility suggested by the negligible penetration of KCl into these cells (Chiarandini, 1976). A very low anionic permeability also is found in frog and avian multiply innervated fibers (Stefani and Steinbach, 1969; Huerta and Stefani, 1975). The time constant of the tonic fibers of rat inferior rectus is 29.8 ± 4.8 msec, which is about nine times larger than in the twitch fibers (D. J. Chiarandini and E. Stefani, unpublished results).

One distinguishing feature of the tonic or MINC fibers is their inability to generate action potentials. This property was recognized in cat by Hess and Pilar (1963) and in rabbit inferior oblique by Matyushkin (1964) and is also present in the slow or tonic fibers of amphibians (Kuffler and Vaughan Williams, 1953a). Hess and Pilar (1963), recording intracellularly from cat superior oblique, found that supramaximal stimulation of the trochlear nerve induced in some fibers complex synaptic potentials which produced maximal depolarizations of 12–14 mV but failed to induce action potentials. They examined a total of 20 fibers with resting potentials ranging from −30 to −70 mV, with an average of −45 mV. Twelve of these fibers were bathed with a Mg^{2+}- and Ca^{2+}-enriched Krebs solution and 8 others were bathed with normal Krebs solution. Action potentials could be evoked with nerve stimulation in only one fiber from the latter group.

Matyushkin (1964) conducted experiments *in vivo* and analyzed two systems: the trochlear nerve–superior oblique muscle of rabbit, and the oculomotor nerve–inferior oblique muscle of cat. He found that stimulation of the nerve induced two types of electrical potentials in the muscle fibers, fast and slow potentials, and recognized the former as action potentials. Using the anodal block technique of Kuffler and Vaughan Williams (1953a), he was able to stimulate selectively the small diameter axons which presumably innervate multiply innervated fibers. He observed with intra- and extracellular recording only slow potentials to which he attributed a synaptic origin, but no action potentials.

The concept that tonic fibers do not give action potentials when stimulated was challenged by Bach-y-Rita and Ito (1966). They were able to record action potentials *in vivo* from fibers of cat superior rectus and inferior oblique which they identified as multiply innervated. However, the validity of the criteria used by these authors to identify the innervation of these fibers has been questioned (Pilar, 1967; Barmack *et al.*, 1971). The existence of multiply innervated fibers with action potentials is of sufficient concern to warrant discussion in a separate section.

Further studies in cat, rabbit, and rat extraocular muscles have definitely confirmed the existence of multiply innervated fibers without action potentials. Pilar (1967) repeated and extended his previous analysis and confirmed that the synaptic potentials evoked by nerve stimulation were of a graded nature and had different time courses and latencies, indicating the presence of multiple innervation in this fiber type. As in his previous study he did not observe action potentials after depolarization of these fibers by intracellularly injected current or by trains of repetitive synaptic potentials.

In rabbit, Ozawa *et al.* (1969) recorded the spontaneous electrical activity of superior rectus simultaneously with intracellular and surface electrodes, recognizing fibers in which more than one source of synaptic potentials could be distinguished on the basis of their different frequency of occurrence and time course. These fibers were identified as multiply innervated, and were thereafter stimulated with intracellularly injected current to trigger action potentials. Subsequently, the fibers were marked with the dye Fast Green FCF and their type of innervation was determined histochemically with cholinesterase staining. From a total of 53 multiply innervated fibers identified electrophysiologically only 8 were successfully marked and retrieved; all were recognized histochemically as being multiply innervated. None of the 53 fibers could be stimulated to produce action potentials. Nevertheless, in another series of similar experiments briefly mentioned in the same article, these authors found that from numerous fibers responding with action potentials to intracellular stimulation, only two were recognized with cholinesterase staining as having multiple nerve endings. These two fibers, however, did not exhibit the normal spontaneous electrical activity seen in the other multiply innervated fibers.

Matyushkin and Drabkina (1970) examined *in vivo* in rabbit superior oblique muscle a very large number of multiply innervated fibers that had very low resting potentials with a mean value of −32 mV. They recorded spontaneous electrical activity similar to that found by Ozawa *et al.* (1969) and did not observe action potentials when the trochlear nerve was stimulated.

Lennerstrand (1974) examined in cat inferior oblique 5 motor units of fibers that contracted after nerve stimulation but in which no action potentials could be detected with extracellular recording. In addition, he recorded electrical activity intracellularly from 3 of these fibers which had a mean resting potential of −52 mV and found that nerve stimulation induced slow synaptic potentials but not action potentials. He concluded that these fibers were similar to the multiply innervated fibers described by Hess and Pilar (1963).

The foregoing studies clearly support the concept that one type of multiply innervated fiber does not generate action potentials upon stimulation. To conclude that this is a genuine property of these muscle fibers it is necessary, however, to rule out the possibilities that the absence of spikes is due to a depolarization or "cathodal depression" of the fibers or to the application of stimulation insufficient to reach electrical threshold. It is clear from the literature that most of the fibers studied had a low resting potential, about −50 mV or even less, as in the experiments of Matyushkin and Drabkina (1970). This level of membrane polarization is less negative than the estimated threshold for the action potentials in the singly innervated fibers of rat extraocular muscles which is about −55 to −60 mV and less negative than the sodium activation in frog skeletal muscle, which is about −55 mV (D. J. Chiarandini and E. Stefani, unpublished results; Costantin, 1968). Therefore, it is possible that in most of the tonic fibers studied, if an action-potential mechanism existed, it was inactivated by depolarization of the cells so that it could not be triggered when nerve or intracellular stimulation was applied.

In rat inferior rectus, D. J. Chiarandini and E. Stefani (unpublished results) studied several multiply innervated fibers all of which had an initial resting potential of −70 mV or more negative, with the exception of one in which the initial resting potential was −60 mV. The fibers were impaled with two microelectrodes and the second penetration was accompanied by a drop of several millivolts in the resting potential. By injecting DC current intracellularly this depolarization could be compensated and the membrane potential set back at −70 or −80 mV. Thereafter, an attempt was made to evoke action potentials by passing depolarizing pulses of different intensities and durations. The fibers failed to produce action potentials even if depolarized to 0 mV (Fig. 4). The only signal evoked by very large depolarizations was a "local response" of still undetermined origin. These experiments demonstrated unquestionably that fully polarized tonic fibers do not produce action potentials. Moreover, Pilar (1967) showed convincingly that some cells with a resting potential of −65 mV did not respond with action potentials when depolarized to about −30 mV by repetitive nerve stimulation (Fig. 4).

Hess and Pilar (1963) carried out a careful analysis of the synaptic potentials recorded in fibers in which no action potentials were observed and demonstrated unequivocally the multiple nature of their innervation. As in the case of singly innervated fibers they observed spontaneous synaptic potentials which were called miniature small nerve junction potentials (msjps), following the nomenclature used by Kuffler and Vaughan Williams (1953a). The msjps ranged in amplitude from baseline noise to about 6 mV, occurred at a low frequency (1–3 per sec) and had an apparent random distribution of intervals. Figure 5 shows this type of potential recorded from cat and rat extraocular muscles.

The msjps have characteristics which easily distinguish them from the

FIG. 4. Failure of tonic fibers to generate action potentials. (Left) Rat inferior rectus. The fiber was identified as multiply innervated by the presence of characteristic miniature small junction potentials. The intracellular application of pulses of current that depolarized the cell from a membrane potential of −70 mV to up to −5 mV failed to evoke an action potential. However, a transient and fast signal appeared at the beginning of the three pulses (arrows). These local responses were followed by a slowly rising depolarization (D.J. Chiarandini and E. Stefani, unpublished results). (Right) Cat superior oblique. The fiber was depolarized by repetitive activation of small junction potentials. Resting potential of the fiber was −65 mV. In spite of a depolarization of about 35 mV, no action potentials were evoked. (From Pilar, 1967 by courtesy of the *Journal of General Physiology*.)

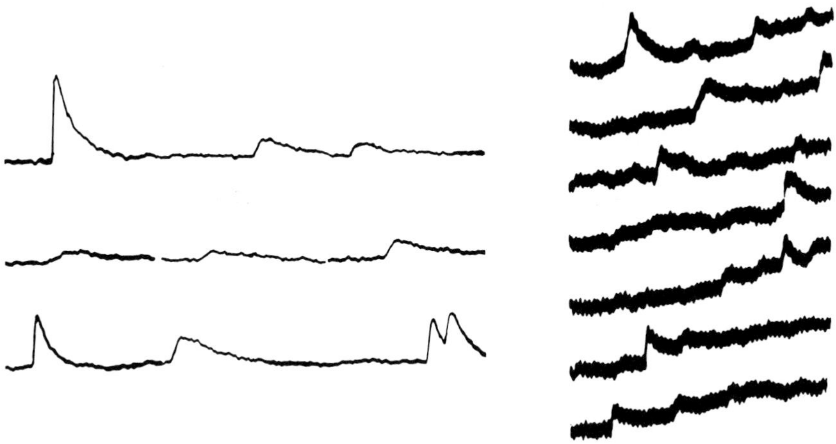

FIG. 5. Miniature small junction potentials in tonic fibers. (Left) Cat superior oblique. Calibration: 1 mV, 10 msec (From Hess and Pilar, 1963). (Right) Rat inferior rectus. Calibration: 1 mV, 200 msec. Notice that in both preparations the rise and half-decay times of the miniature small junction potentials have a wide range of variation. (D. J. Chiarandini and E. Stefani, unpublished results.)

mepps, and which can be used for immediate identification of multiple innervation of the fiber in which they occur. They can be recorded from any region along the muscle length without being limited to the site of the nerve entrance and branching, as is the case for the mepps. They have a longer duration than the mepps. In rat the duration is about 10 times longer (Figs. 3 and 5) with longer rise and half-decay times. Although rise and half-decay times of the msjps have a wide range of values (Table III), slow and small potentials prevail (Fig. 5), and the distribution of the amplitudes of the msjps is skewed with a predominance of small voltages (Fig. 6) (Hess and Pilar, 1963; D. J. Chiarandini and E. Stefani, unpublished results).

The large variation in amplitude and in time course of the msjps and the skewed distribution of their amplitudes is readily explained by the multiple innervation of the fibers. There should be more potentials originating at some distance from the recording micropipette than near it. Other parameters being equal, the former should be smaller and slower than potentials arising from junctions near the recording site because of attenuation in amplitude and in rates of rise and decay imposed by the cable properties of the muscle fibers. A similar situation exists in the tonic fibers of amphibia (Kuffler and Vaughan Williams, 1953a; Burke, 1957). However, amplitude and rise time of msjps are not in strict correlation and potentials of very similar amplitude may have very different rise times (Fig. 5). This suggests that the size of the potentials is not the same at all the synaptic junctions.

TABLE III

Characteristics of Miniature Small Junction Potentials

Preparation	Voltage (mV)	Frequency (sec^{-1})	Rise time (msec)	Half-decay time (msec)	
Cat superior oblique *in vitro*, 20°–25°C	0.2–6.0	1.2–2.5	2.0–24.0	5.0–35.5	Hess and Pilar (1963)
Rat inferior rectus *in vitro*, 20°–25°C	0.3–4.8	0.9–3.1	7.5–90.0	7.0–80.0	D. J. Chiarandini and E. Stefani (unpublished results)
Frog iliofibularis *in vitro*, 15°–22°C	0.25–2.5	0.03–4.0			Burke (1957)

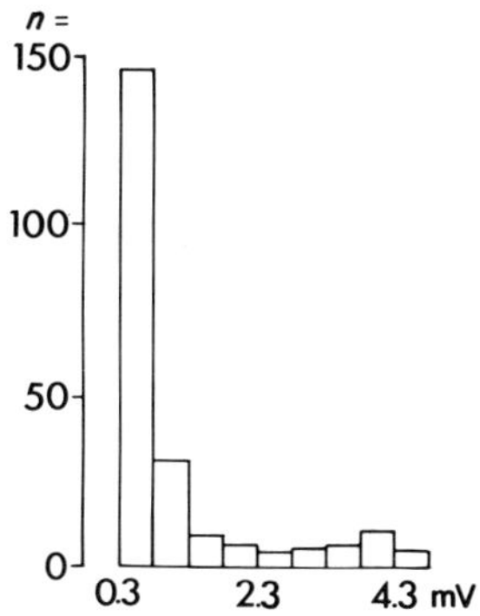

FIG. 6. Histogram of the amplitude distribution of the miniature small junction potentials in rat extraocular muscles. The distribution is skewed with a predominance of small-amplitude potentials. (D. J. Chiarandini and E. Stefani, unpublished results.)

Multiply innervated fibers have another type of synaptic potential called small nerve junction potentials (sjps), which are physiologically equivalent to the end plate potentials described in singly innervated fibers. Small nerve junction potentials are produced when an action potential, either induced by a stimulus applied to the nerve or generated by a neuron of the oculomotor nuclei, reaches the nerve ending and evokes the simultaneous release of many quanta of ACh. The sjps were originally described by Kuffler and Gerard (1947) in the multiply innervated muscle fibers of frogs and their name is derived from the observation that they are evoked by the stimulation of small-diameter motor nerves. Hess and Pilar (1963) were the first to describe this type of potentials in extraocular muscles and adopted for them the name given by Kuffler and Gerard (1947).

Hess and Pilar (1963) and Pilar (1967) demonstrated that single, supramaximal nerve stimulation yields sjps with amplitudes of no more than 15 mV and which have various components with different amplitudes, time courses and latencies. The sjps can be recorded anywhere along the muscle length and the number of unitary components evoked can be varied by modifying the strength of stimulation of the motor nerve. An example is shown in Fig. 7 in which the records were taken in sequence as the stimulus strength was gradually increased from just above threshold to maximal. Increments in intensity resulted in more complex sjps and the sjp obtained with maximal stimulation (Fig. 7) was the summation of numerous unitary potentials. Discharges of such unitary components occur spontaneously in animals with intact innervation and will be discussed later.

As stressed by Hess and Pilar (1963), the presence in the sjps of numerous unitary potentials with different amplitudes, time courses, and laten-

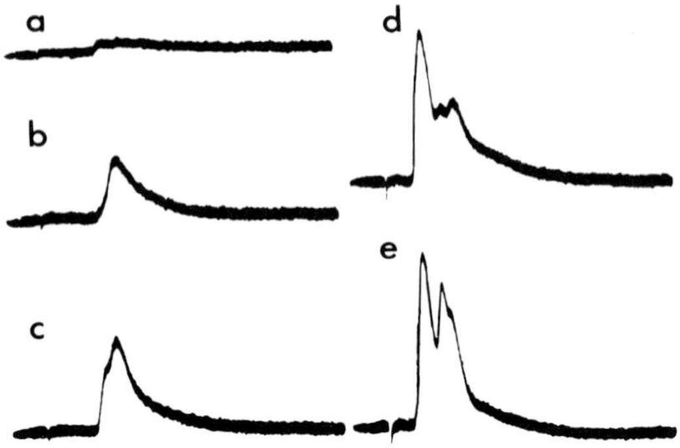

FIG. 7. Intracellularly recorded small junction potentials in cat superior oblique. The application of stimuli of increasing intensity from just above threshold (a) to maximal (e) evoked small junction potentials of increasing complexity which resulted from the summation of numerous components of varying latency, amplitude and time course. (From Hess and Pilar, 1963.)

cies indicates that there are many areas of innervation in the muscle that are at varying distances from the recording site and some that are closely spaced with respect to the dimensions of the space constant. Moreover, they pointed out that the differences in thresholds and latencies of the various components of the sjps indicate that the multiple innervation is polyneuronal, i.e., several motor axons with different thresholds and conduction velocities innervate the muscle. Presumably, this difference in the excitability of the axons is the result of their different diameters. In rat inferior rectus, sjps similar to those described in cat have been recorded (D. J. Chiarandini and E. Stefani, unpublished results). Polyneuronal innervation is also found in multiply innervated fibers of amphibian and avian muscles (Kuffler and Vaughan Williams, 1953a; Ginsborg, 1960).

In rabbits in which the muscle innervation was kept intact, intra- and extracellular recordings have shown the existence of one type of fiber that exhibits spontaneously a continuous discharge of slow depolarizing potentials, presumably evoked by the firing of neurons (Ozawa, 1964; Nemet and Miller, 1968; Ozawa *et al.*, 1969; Matyushkin and Drabkina, 1970; Schulze, 1977). In the same fiber usually up to 4 slow potentials could be distinguished on the basis of their different amplitudes and time courses. Each of these distinct potentials appeared with a regular frequency, which ranged from 7 to 80 per sec. Thus, the electrical activity consisted of the superimposed discharge of several different slow potentials. The amplitude and duration of the slow potentials varied from fiber to fiber and ranged from 1.5 to 26 mV and from 5 to 30 msec. The slow potentials with different amplitudes and time courses are thought to originate from separate neuromuscular junctions so that their variation is a consequence of dissimilar distances between the synaptic regions, where they originate, and the recording microelectrode. The slow potentials can be considered as being identical to the unitary potentials which, when summated, make up the sjps evoked by the nerve stimulation discussed previously.

Ozawa (1964) classified the spontaneous activity of rabbit tonic fibers according to the amplitude of the signal. He separated them into slow potentials (10–20 mV) and giant potentials (20–50 mV). The two types occurred at regular frequencies, independently or in a mixed fashion. He considered that slow potentials were of synaptic origin and that giant potentials represented abortive action potentials without overshoot. However, this conclusion is disputable. Giant potentials could be considered abortive action potentials if their main difference from action potentials were the absence of overshoot, a possible consequence of the somewhat low resting potential of the fibers studied by Ozawa. However, the examination of Ozawa's record shows that the duration of the giant potentials is at least five times that of the action potentials that he recorded in twitch fibers. Such dissimilarity cannot be accounted for by a difference of several millivolts in the resting potential of both types of cells. An alternative explanation can be suggested for the giant potentials. Large slow potentials could depolarize the fibers to a certain level of membrane potential and thus trigger a "local response" with fast rise and decay times. In multiply innervated fibers of rat it has been shown that large depolarizations evoked a local response of this kind (Fig. 4).

Matyushkin and Drabkina (1970) analyzed the amplitude and time course of slow potentials in rabbit tonic fibers and on the basis of their amplitudes distinguished two types: principal or large and additional or small potentials. Applying the cable equation derived by Hodgkin and Rushton (1946) they determined the half-decay time of the principal potentials at their site of origin, and subdivided them into two groups characterized by having a fast and slow decay. They concluded that the mechanism involved in the decay of the two types of potentials was different and called the fast-declining potentials, which can be as large as 26 mV, "spikelike potentials." Some comments are pertinent to this conclusion that inititally may seem to strengthen Ozawa's view. The fibers studied by Matyushkin and Drabkina (1970) had very low resting potentials with a mean value of −32 mV, at which the Na^+ activation mechanism of the action potential should have been inactivated. This should rule out the possibility that any electrical activity recorded by them involved the spike mechanism. Their quantitative analysis of the rate of decay of the principal potentials is based on the erroneous assumption that the cable equations derived by Hodgkin and Rushton (1946) can be applied to the multiply innervated fibers of rabbit extraocular muscles. The studies performed in cat and rat strongly suggest that multiply innervated fibers have very high values of input resistance and consequently of space constant, therefore, they should behave electrically as short cables and not as infinite cables, the boundary condition for which Hodgkin and Rushton (1946) have computed the cable equation.

3. Multiply Innervated Fibers with Action Potentials or MIC Fibers

As has been mentioned, Bach-y-Rita and Ito (1966) reported in cat extraocular muscles the presence of fibers that are multiply innervated and that, in contrast with the previous findings of Hess and Pilar (1963) and Matyushkin (1964), are capable of producing propagated action potentials. For this reason they named these muscle cells "slow multi-innervated twitch fibers." The reported number of these fibers was rather large and represented about one-third of the fibers studied by the authors.

Bach-y-Rita and Ito (1966) identified the multiply innervated twitch fibers by their response to anodal block stimulation, by their low resting potential, and by the characteristics of their spontaneous synaptic activity. The anodal block stimulation technique was developed by Kuffler and Vaughan Williams (1953a) and Burke and Ginsborg (1956) to stimulate selectively in frog the slow conducting motor axons that are considered to be the main source of innervation of the multiply innervated fibers in frog-limb and mammalian-extraocular muscles (Kuffler and Vaughan Williams, 1953a; Matyushkin, 1964). This technique consists of stimulating electrically the nerve using a pair of platinum wire electrodes of which the anode is placed closer to the muscle. During the application of square pulses the nerve fibers are simultaneously depolarized and hyperpolarized at the sites of the cathode and anode, respectively. The cathodal stimulation generates an action potential that, to reach the muscle, has to traverse the hyperpolarized region of the nerve fiber at the site of the anode. If, at the time that the action potential arrives at the anodal region, the hyperpolarization is large enough, nerve conduction will be blocked. By adjusting the interelectrode distance and the pulse duration and intensity, it is possible to eliminate the action-potential conduction of the fast conducting axons without affecting the action-potential conduction of the slow conducting nerve fibers. Bach-y-Rita and Ito (1966) applied this technique to the oculomotor nerve-inferior oblique system *in vivo* using slowly decaying, saw-tooth pulses with a time constant of 30 msec, repeated at different frequencies. After sectioning all the nerve branches except the distal branch, they showed with extracellular recording that cathodal or conventional stimulation evoked in the muscle both fast and slow action potentials, whereas anodal block stimulation evoked only slow action potentials.

Intracellular recording from numerous fibers of cat inferior oblique demonstrated that only some of them responded with an action potential to the anodal block stimulation (Bach-y-Rita and Ito, 1966). The resting potential of the activated fibers ranged between −20 and −65 mV and only in a fraction of them with resting potentials of −50 mV or more negative did the action potential have an overshoot. The majority of the cells exhibited nonovershooting depolarizations. Bach-y-Rita and Ito (1966) studied further the electrical properties of this group of fibers using single microelectrodes connected to a bridge circuit or double barreled microelectrodes. In 45 fibers they measured an input resistance of 0.9 ± 0.46 MΩ and a time constant of 10.2 ± 1.46 $\mu F/cm^2$. Assuming a fiber diameter of from 10 to 15 μm, they calculated a specific membrane resistance of about 4,500 Ω-cm^2 and a specific membrane capacity of 2.48 $\mu F/cm^2$.

Bach-y-Rita and Ito (1966) studied in the MIC, or multiply innervated fibers with action potentials, the spontaneous synaptic potentials which they named mepps to distinguish them from the msjps described previously by Hess and Pilar (1963) in the MINC or tonic fibers. These authors found that mepps occur at random, at frequencies between 0.8 and 5.2 per sec and that they could be recorded in different sites along the length of the fibers. Their amplitudes ranged from about 0.2 mV to 4.3 mV and had a skewed distribution with two populations. The largest had a peak of about 0.5 mV while the smallest encompassed mostly potentials from 1.0 to 3.0 mV and had no obvious peak. The amplitude and time course of MIC fiber mepps varied much more than those of twitch fiber mepps. The means for the rise and half-decay times of the mepps smaller than 1.5 mV were 0.97 msec and 2.85 msec, respectively. Mepps larger than 1.5 mV had a shorter time course than smaller mepps. The overall range of variation of the rise and half-decay times, important criteria for the recognition of multiple innervation of muscle fibers (Kuffler and Vaughan Williams, 1953a; Burke, 1957; Ginsborg, 1960; Hess and Pilar, 1963), have not been reported by Bach-y-Rita and Ito. The synaptic activity evoked by stimulation of motorneurons or nerves was not studied by these authors in the MIC fibers.

Bach-y-Rita and Ito (1966) originally thought that each of the multiply innervated fibers with action potentials received innervation from more than one neuron, but in a very recent study Bach-y-Rita and Lennerstrand (1975) concluded that polyneuronal innervation was not present in these fibers. They analyzed this question by studying tension excess or deficit, post-tetanic potentiation and cross fatigue of twitches evoked in lateral rectus muscles. Polyneuronal innervation, however, exists in tonic fibers as mentioned before.

Prompted by the work of Bach-y-Rita and Ito (1966), Pilar (1967) reexamined the electrical properties of multiply innervated fibers of cat superior oblique and concluded that *in vitro* and *in vivo* these fibers did not generate action potentials. Combining the recording of tension and extracellular electrical activity he showed further that there was a component of tension that was sustained and did not involve propagation of impulses. He attributed the origin of this tension to tonic or multiply innervated fibers without action potentials. Pilar criticized the criteria used by Bach-y-Rita and Ito (1966) for distinguishing multiply innervated fibers. For instance, he pointed out that it was incorrect to assume that a fiber was multiply innervated because it had a resting potential lower than the singly innervated or twitch fibers since, as mentioned by Bach-y-Rita and Ito (1966), resting potentials of the two types of fibers showed considerable overlap. The fibers that were excited by anodal stimulation had resting potentials between −20 and −65 mV while the fibers that did not respond to anodal block had values between −20 and −110 mV. Furthermore, Pilar (1967) indicated that anodal block stimulation had been used without the necessary monitoring of nerve responses to determine whether only small nerve fibers were being stimulated selectively.

Lennerstrand (1974) studied in cat inferior oblique the electrical responses of single motor units to nerve stimulation with extracellular recording *in vivo*. He explored the surface of the muscle with a bipolar electrode recording systematically from 10 to 12 sites distributed along 12–15 mm of the muscle length. In this way he located end plate regions from the position of the electrode at which the evoked extracellular action potential reversed polarity and had the briefest latency.

In 16 of the motor units studied by Lennerstrand (1974) only one end plate could be detected, which indicated that the fibers were singly innervated. In 14 other units 2 or more end plate regions were found. The area of innervation of these fibers occupied from 5 to 8 mm of the muscle length and was absent in the peripheral part of the motor units. The mean conduction velocity in the muscle fibers with several end plates was 1.72 m/sec, which was slower than the conduction velocity of 2.93 m/sec recorded in the singly innervated fibers in the same study. Lennerstrand (1974) ruled out the possibility that the multiple end plate regions identified with the extracellular electrode were located in adjacent singly innervated fibers with staggered end plates, on the basis that in singly innervated fibers of other eye muscles the end plates are restricted to a narrow band of about 2 mm length. However, as has been pointed out by Browne (1976), Cooper and Fillenz (1955) found in the same muscle used by Lennerstrand, that the end plates formed a wide scattered region. Moreover, Mayr *et al.* (1975) reported that, in the global layer of cat lateral rectus and other oculorotatory muscles, there were present several bands of innervation which presumably arise from the existence of short singly innervated fibers that do not run from end to end of the muscle but are arranged in series and are apparently interconnected by myomyous junctions. These anatomical findings cast doubts about the real identity of the fibers that Lennerstrand recognized as multiply innervated with action potentials.

In the reports of Bach-y-Rita and Ito (1966) and Lennerstrand (1974), the identification of multiply innervated fibers with action potentials has been made on the basis of indirect evidence. None of these authors obtained direct proof of the existence of MIC fibers by examining in the same muscle cell both the characteristics of its synaptic activity and its ability to give action potentials. The observation in a given fiber of spontaneous synaptic potentials that have a wide range of rise and half-decay times would be enough to demonstrate that the fiber is multiply innervated.

Multiply innervated fibers with action potentials have been identified directly using different techniques by Hess and Pilar (1963) and by Ozawa *et al.* (1969), but in both studies, these fibers appear to be an extremely small percentage of the total muscle cell population. This finding contrasts dramatically with the postulation that MIC fibers correspond to about one-third to one-half of all the fibers studied by Bach-y-Rita and Ito (1966) and Lennerstrand (1974). It should be remembered, however, that in these studies different regions of the muscles were explored and it is conceivable that different regions may have an uneven number of MIC fibers. This could explain, at least partially, the discrepancy found in the literature with respect to the size of the MIC-fiber population. Therefore, further experiments aimed at identifying directly multiply innervated fibers with action potentials in different regions of the muscle should provide a definite answer to the question of how large is the population of MIC fibers in extraocular muscles.

C. Contractile Properties of EOM Fibers

The extensive research into the electrical properties of eye muscles that has been discussed in the preceding section shows that these muscles contain fibers in which the arrival of the nerve impulse at the neuromuscular junction induces in the muscle cell either an action potential or a small junction potential (sjp). In turn, these electrical signals evoke the development of two types of contractile responses, either an all or none, transient contraction or twitch, or a graded, slow, tonic tension. Thus, two types of contractile systems can be recognized in eye muscles, one consisting of fibers that generate twitches and the other consisting of fibers that develop tonic tension.

1. Twitch or Phasic Motor System

The dominant cellular element of this system is the singly innervated or twitch fiber which, as in all other twitch muscle fibers of vertebrates, is activated by action potentials to contract. It is generally accepted that this type of fiber is innervated by a single axon with a large diameter and a low threshold for cathodal stimulation.

The contractile properties of the twitch motor system have been defined by considering the time taken for the contraction to reach a maximal value and to relax, and by the frequency of stimulation at which fused tetanus is observed. Usually, the first parameter is expressed as the time from the onset of the contraction to its peak value, or contraction time, and the second parameter is expressed as the time necessary for the tension to decline to half of its maximal value, or half-decay time. The characteristics of the twitch motor system in several extraocular and limb muscles of various mammals are shown in Table IV. Both contraction and half-decay times of the different extraocular muscles are remarkably similar with mean values of about 7 and 8 msec, respectively. The fusion frequency of the tetanus for whole muscles shows instead a wider range of values, from 120 to 450 pulses/sec. This variability might be due to differences in the frequency response and sensitivity of the transducer and recording systems used in the diverse studies. A comparison of the results obtained in extraocular and limb muscles shows clearly how fast the contraction–relaxation process takes place in the first group of muscles. Eye muscles reach peak tension and relax 4–12 times faster than limb muscles.

Eye muscles have, in addition, other remarkable properties. The ratio between the twitch and tetanic tension is small, about 0.1; at least two times smaller than in limb muscle. Therefore, the tension developed by the muscles during a train of impulses may be up to ten times that of the twitch evoked by a single action potential. Barmack (1977) has pointed out that this characteristic favors the postulation that frequency modulation is a mechanism for controlling the magnitude of force developed by the muscle. Close and Luff (1974) demonstrated that the tetanic tension generated by rat inferior rectus per cross sectional area of muscle is about 1 kg/cm^2 and equivalent to only about one-third to one-half of the values reported for rat limb muscles (Close, 1969). Similar measurements for eye muscles of other species are not available.

As shown in Table IV the fusion frequency reported for most of the extraocular muscles is about 350–400 per sec. and it is known that at these frequencies of stimulation the muscles develop maximal tetanic tensions (Cooper and Eccles, 1930; Barmack *et al.*, 1971; Fuchs and Luschei, 1971a; Lennerstrand, 1974). However, electrophysiological studies have demonstrated that about 20% of the nerve cells of the motor nuclei fire at rates of 600 per sec or above (Fuchs and Luschei, 1970; Robinson, 1970; Schiller, 1970). This observation raised the question of why a motor neuron would fire at such high frequencies, and it was postulated that, as in other muscles, faster motorneuron firing rates could result in a more rapid

TABLE IV. Contractile Characteristics of Twitch Muscle Fibers

	Contraction time (msec)	Half-decay time (msec)	Fusion frequency (pulses/sec)	Ratio of twitch tension to tetanus tension	
Extraocular muscles					
Cat					
Lateral rectus	7.5		350	0.09	Cooper and Eccles (1930)
Lateral rectus	6.4	8.2	>250	0.13	Barmark *et al.* (1971)
Inferior oblique					
Fast motor units	5.7–6.9	7–8	300–450	0.09	Bach-y-Rita and Ito (1966)
Slow motor units	20–27	38–44	25		
Inferior oblique					
Fast motor units	3.5		>200	0.07	Lennerstrand (1974)
Slow motor units	20		<200	0.07	
Inferior oblique	5.6–8.1	10.7–15.5	275–425		Hanson and Lennerstrand (1977)
Rat					
Inferior rectus (*in vitro*, 35°C)	5.3	5.6		0.13	Close and Luff (1974)
Inferior oblique	6.4–8.5	5.4–8.7	250–350		Hanson and Lennerstrand (1977)
Rabbit					
Superior oblique	8–10	10	>120		Matyushkin (1964)
Monkey					
Lateral rectus	7.0	6.4	400		Fuchs and Luschei (1971a)
Limb muscles					
Cat					
Soleus	105		32	0.29	Cooper and Eccles (1930)
EDL	31		110	0.27	Cooper and Eccles (1930)
Triceps surae					
Fast motor units	≤30	≤30	85	0.38	Burke (1967)
Slow motor units	≥40	≥50	25	0.19	Burke (1967)
Rat					
EDL	11.8	7.7		0.20	Close (1967)
Soleus	36.0	51.0		0.20	Close (1967)

development of tetanic tension (Fuchs and Luschei, 1970a; Robinson, 1970). This question was examined in cat and monkey lateral rectus by Barmack *et al.* (1971) and Fuchs and Luschei (1971a) who found that stimulation frequencies between 400 and 600 per second did not induce additional tetanic tension but increased the rate of rise of the tetanus by 20–25% (Fig. 8). This property results in a better transient response of the control system of ocular movements (Barmack *et al.*, 1971).

The existence in mammalian skeletal muscle of different types of twitch fibers has been recognized at the structural and physiological level for many years. For instance, three main types of fibers can be distinguished histochemically by the degree of activity of ATPase at pH 9.4 and of the enzymes related to glycolysis and to mitochondrial oxidative processes (Stein and Padykula, 1962; Yellin and Guth, 1970; Barnard *et al.*, 1971). The analysis of the contractile properties of different muscles has demonstrated the existence of two types of twitch muscle fibers, fast and slow. The investigation of the twitch characteristics of single motor units has reaffirmed these concepts (Burke, 1967; Close, 1972). The various types of singly innervated fibers recognized morphologically in extraocular muscles strongly suggest the existence in these muscles of more than one type of twitch fibers. However, investigations of this aspect of eye-muscle contractility carried out by several workers have yielded contradictory results.

Bach-y-Rita and Ito (1966), utilizing the anodal block technique, identified in cat inferior oblique the existence of fibers that had relatively slow

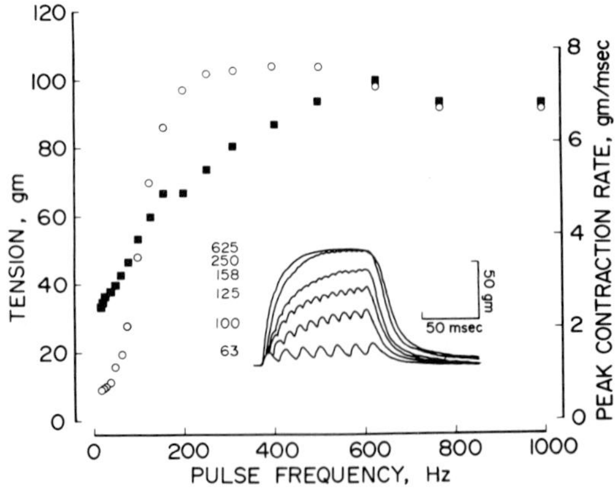

FIG. 8. Effect of stimulus frequency on isometric tension and rate of tension development. Circles, peak tension. Squares, peak rate of tension. The stimulus was a 100 msec train of 0.1 msec, 0.6 mA pulses at different frequencies delivered to the VI nerve. (From Barmack *et al.*, 1971 by courtesy of the *Journal of Neurophysiology*.)

twitches and a fusion frequency of 25 per sec. The contraction and relaxation times of these fibers were longer than those of the twitches observed by these and other authors using conventional cathodal stimulation (Table IV). The slow-contracting fibers were identified as multiply innervated with action potentials (i.e., MIC fibers) on the basis of their response to anodal block stimulation. Barmack *et al.* (1971) have questioned the specificity of the anodal block stimulation used in these experiments because with the parameters employed, 30 msec pulses applied at a frequency of 25 per sec, stimulation approached a direct current condition. Consequently, the strength of the cathodal and anodal currents required to increase and decrease nerve excitability could vary and even reverse in sign.

Further evidence supporting the concept of slow-twitch fibers in cat inferior oblique is found in the work of Lennerstrand (1974), Lennerstrand and Bach-y-Rita (1974) and Golberg *et al.* (1976). Lennerstrand (1974) isolated single motor units by stimulating with threshold intensities one of the three branches of the inferior oblique nerve of cat. With this technique he analyzed the properties of about 70 units, some of which already had been identified with extracellular recording as being singly or multiply innervated. He found that the motor units could be divided into fast and slow units according to contraction time and fusion frequency (Fig. 9 and

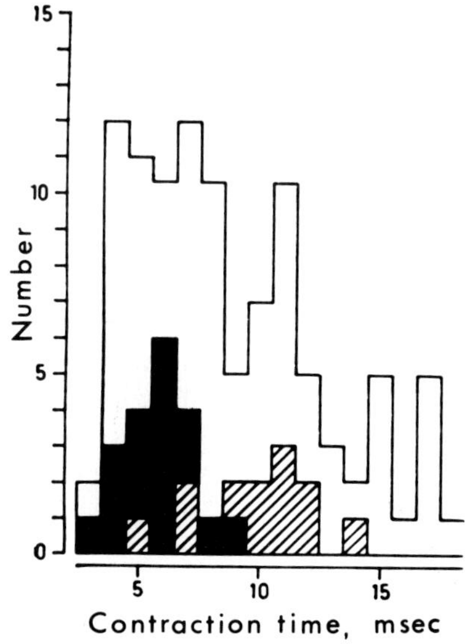

FIG. 9. Distribution of twitch contraction time of single motor units. Cat inferior oblique. Some of the units were identified electrophysiologically as singly innervated (SI) fibers (filled bars) or multiply innervated with action potentials or MIC fiber (striped bars). (From Lennerstrand, 1974 by courtesy of *Acta Physiologica Scandinavica.*)

Table IV). Most of the slow-twitch fibers were considered multiply innervated with action potentials by Lennerstrand (1974), but this conclusion should be regarded only as tentative as was discussed in relation to the electrical properties of MIC fibers.

The contraction time of the motor units varied over a wide range from 3.5 to 20 msec and had a bimodal distribution. The motor units recognized as singly innervated reached peak twitch tension in 3.5–9 msec, while the motor units considered to be multiply innervated reached peak twitch tension in 5–15 msec. The fusion frequency of the tetanus also had a bimodal distribution and its value was chosen to classify the contractile properties of the motor unit as fast or slow. Fast units, made up of singly innervated fibers, had a fusion frequency of 200–350 stimuli per sec. The slow motor units, which according to Lennerstrand (1974) contained mostly MIC fibers, had a fusion frequency below 200 per sec. Fast units usually had higher values of tetanic tension than slow units, but as in the case of the other parameters, there was considerable overlap. Experiments designed to study the fatigability of the two types of motor units under repeated stimulation for 30 sec demonstrated that, in general, the slow units were more resistant to fatigue than the fast units, but even at stimulation rates of 300 per sec, 75% of the units of the fast type were able to retain 30% of their initial tension. The few units that fatigued completely were fast ones (Lennerstrand, 1974).

In cat inferior oblique Lennerstrand and Bach-y-Rita (1974), recording the electromyogram and the tension, showed that threshold stimulation of the nerve evoked a long latency biphasic potential and a twitch with a mean contraction time of 11 msec. Further increases in the intensity of stimulation brought about a short latency potential and a decrease in the contraction time to about 5 msec, indicating the recruitment of other faster fibers and thus suggesting the presence of two populations of twitch fibers. Since the axons with the lowest threshold of stimulation have the largest diameters, these experiments suggest that, contrary to what is expected, some slow motor units appear to be innervated by large, fast conducting axons. Goldberg *et al.* (1976) activated motor units in cat lateral rectus by stimulating directly motorneurons of the abducens nucleus with intracellular injections of current according to the technique of Burke (1967). They confirmed the existence of fast- and slow-twitch motor units and verified that there is no firm relationship between motor unit contraction time and conduction time of the nerve fibers. Fast- and some slow-contracting motor units appeared to be innervated by fast-conducting axons. That slow motor units may receive innervation from large axons was originally suggested by Nemet and Miller (1968).

The existence of slow-contracting fibers in cat extraocular muscles with a fusion frequency of about 25 per sec, as reported by Bach-y-Rita and Ito (1966), was not confirmed by Barmack *et al.* (1971) and by Fuchs and Luschei (1971a). The first authors stimulated cat lateral rectus with two successive pulses of various intensities and found that the amplitude of the second twitch remained unchanged if the interval between pulses was 25 msec or more. Similarly, using trains of pulses, they did not observe an increment in the tension until the frequency of the stimuli exceeded 40 per sec. It would have been expected that these patterns of stimulation would demonstrate the summation and, eventually, the fusion of the slow twitches, since according to Bach-y-Rita and Ito (1966) their contraction and half-decay times are about 25 and 40 msec. Moreover, the examination of the relationship between the size of the twitch response and the contraction and half-decay times showed that as the twitch increased from threshold to maximal amplitude, both the contraction and half-decay times decreased from 8 to 6.4 msec and from 18 to 7.5 msec, respectively. Barmack *et al.* (1971) concluded that only fast-twitch fibers were present in lateral rectus of cat since the increase in strength of stimulation failed to recruit the slow-twitch fiber population which is presumed to have a high stimulation threshold. The gradual decrease of contraction and half-decay times that accompanied the increments in twitch size were attributed to mechanical reasons. With increments in contraction amplitude, the series elastic component should have relatively less damping effect on those two parameters (Collins, 1971). The time courses of the twitches evoked by Barmack *et al.* (1971) with different intensities of stimulation show that the activated fibers were of the fast kind. It is unclear why these authors did not observe the changes in time course of the twitch as reported by Lennerstrand and Bach-y-Rita (1974).

Close and Luff (1974) studied motor units in rat inferior rectus *in vitro* at 35°C. They identified single motor units by using threshold stimulation of small filaments of the oculomotor nerve or by analyzing increments in the records of the summed responses of up to 5 units which had relatively constant differences in threshold. These experiments showed that the contraction and half-relaxation times had a narrow range of values, about half of that of the extensor digitorum longus muscle (Table IV). Using stimulation of increasing intensity they demonstrated, as had Barmack *et al.* (1971), that increments in twitch amplitude were associated with a reduction of about 10% in the contraction time and about 25% in the half-decline time. Since the range of stimulation intensity presumably covered the whole range of axon thresholds, they concluded that slow-twitch fibers, as described by Bach-y-Rita and Ito (1966), are not present in rat inferior rectus. Moreover, in another series of experiments they found that the twitch amplitude was about 1.8 times greater when the temperature in the bathing fluid was lowered to 21°–22°C and that the twitch of curarized muscles showed post-tetanic potentiation when directly stimulated. This observation and the negative correlation between contraction amplitude and temperature are characteristic of limb muscle fast-twitch fibers and are not observed in slow-twitch fibers (Close and Hoh, 1968).

The recent work of Barmack (1977, and personal communication) provides insight about how the contraction of twitch fibers may integrate into an eye movement. This author recorded in rabbit inferior rectus the extracellular electrical activity of muscle fibers activated by a vestibular ocular reflex. He found that there was a linear relationship between motor unit size and their recruitment threshold. The entire motor unit population was recruited over a range of 6 gm, which corresponds to an eye deviation of 12 degrees downward. At levels of isometric tension above recruitment threshold, increments of frequency of firing were followed by increments of force, until a level of saturation was reached. Small motor units have a steeper frequency–force relationship than larger units, they are scattered throughout the muscle but most densely concentrated in the orbital layer. Units generating active force of less than 1 gram were located exclusively in this region. Barmack (1977) pointed out that the lower recruitment threshold and the steeper suprathreshold frequency–force relationship of small motor units may confer a greater precision of oculomotor control when the eye is in primary position with only the smallest units active. Since, normally, the tension developed by the motor units is isotonic, the larger motor units are recruited at shorter muscle lengths. Because of the length–tension characteristic of the muscle, it is expected that at shorter sarcomere lengths, each individual fiber would produce less tension and it is conceivable that this is compensated for by the large size of the units.

2. Tonic Motor System

The work of Matyushkin (1961, 1964) and Hess and Pilar (1963) demonstrated in rabbit and cat eye muscles the presence of a population of muscle fibers that had contractile characteristics similar to those of the tonic or slow fibers of amphibians (Kuffler and Vaughan Williams, 1953b). This contractile system is made up of the tonic or MINC fibers described previously. These fibers are characterized by a relatively slow rate of contraction and relaxation. They respond to a single electric shock with a weak tension, requiring repetitive stimulation to recruit their full contractile response.

Matyushkin (1964) stimulated selectively with the anodal block technique small diameter axons that innervate the tonic fibers of rabbit and cat eye muscles. He recorded a slow rising tension that increased in amplitude with increments in the stimulation frequency from 20 to 120 per sec, although the rate of rise remained unchanged. Tensions induced with pulses delivered at a rate of 100 per sec reached a constant value after approximately 0.7 sec of stimulation. The maximal tension developed by the tonic system was obtained with a frequency of 120 per sec or more and it was equivalent to about 10% of the tension developed by the muscle under maximal cathodal stimulation. Matyushkin (1964) indicated that not all the tonic fibers appeared to be excited in these experiments and suggested that the real maximum force of the tonic system could be greater.

In cat superior oblique, Hess and Pilar (1963) were able to separate the phasic from the tonic systems by taking advantage of the different locations of neuromuscular junctions in the fibers of these two systems. By exposing progressively larger portions of a vertically mounted muscle to a saline containing D-tybocurarine, they curarized the junctions of almost all the twitch fibers and left unblocked the transmission at a considerable number of nerve endings of the tonic fibers. The upper record (A) of Fig. 10 shows the partially fused tetanus evoked by nerve stimulation at a frequency of 30 per sec before curarization. Exposure to D-tubocurarine of the

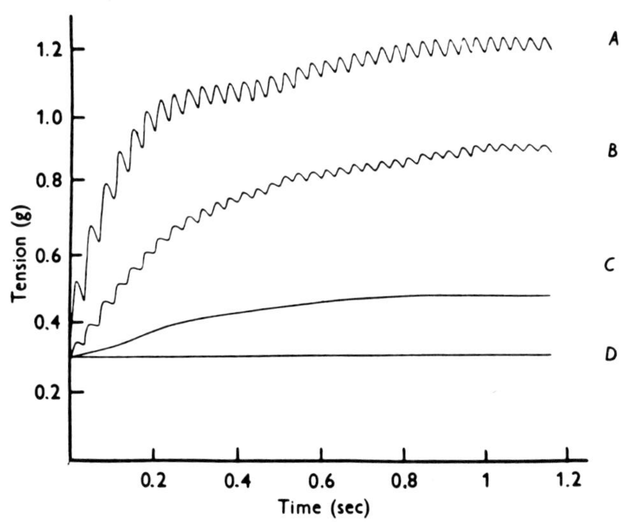

FIG. 10. Tension changes produced by maximal stimulation of the nerve at 30 per sec. Cat superior oblique. (A) Control, partially fused tetanus. (B) Proximal one-fifth of the muscle was exposed to D-tubocurarine and the partially fused tetanus was reduced in amplitude. (C) Curarization of the proximal two-thirds of the muscle, a smooth tension response was obtained with no sign of contractile increments at the stimulation frequency. (D) Full exposure of the muscle to D-tubocurarine. D-tubocurarine 10 μg/ml was applied for 20 min at each level. (From Hess and Pilar, 1963.)

proximal one fifth of the muscle (B) resulted in a reduction of tetanus amplitude. After curarizing the proximal two-thirds of the muscle (C) the nonfused tetanus disappeared and was replaced by a smooth tension in which no sign of the contraction increments seen in records (A) and (B) was observed. Total curarization of the muscle suppressed the response to nerve stimulation (D). The fibers giving distinct tension increments at the stimulation rate of 30 per sec had neuromuscular junctions located at the proximal end of the muscle and were recognized by Hess and Pilar as twitch or singly innervated fibers. The smooth tension which developed when all but the distal one-third of the muscle was curarized has been attributed to the tonic or multiply innervated fibers. Hess and Pilar pointed out that since the contraction of tonic fibers is local, the tension observed in record (C) corresponded to a partial activation of tonic fibers which under normal conditions would produce a tension several times larger.

An elegant demonstration of the existence of twitch and tonic motor systems in cat superior oblique has been provided by Pilar (1967) who recorded simultaneously the electrical activity of the muscle and the tension during a prolonged tetanic stimulation applied to the trochlear nerve. He showed that the onset of a tetanus evoked at 180 stimuli per sec, biphasic action potentials were recorded extracellularly from the muscle and tension rose rapidly to a maximum in about 2 sec. Since the stimulating frequency was lower than the fusion frequency, the tetanus was only partially fused and each stimulus produced a transient tension increase. Each transient increase in tension was accompanied by a propagated action potential in the muscle. After a few seconds, tension began to decline and action potentials appeared with a longer latency and had a smaller amplitude. Some seconds later the partially fused tetanic tension declined to about one-third of the maximal value and became smooth. At the same time the biphasic action potentials disappeared and only monophasic responses with a slower time course were recorded. These experiments clearly confirmed the existence of two contractile systems which have different frequencies of fusion and in which tension is evoked by two kinds of electrical signals. At the onset of stimulation both systems were activated and the tension recording showed transient tension increments, each one accompanied by an action potential. The contractile system underlying these responses was the twitch or phasic system. When the action potentials disappeared because of the fatigue of neuromuscular transmission, only the smooth tension produced by the tonic or slow contractile system remained. The electrical activity that evoked this tension was of the nonpropagated type and corresponded to the sjps.

Studying the motor units of rat inferior rectus, Close and Luff (1974) found that relaxation of the tetanic tension evoked by indirect stimulation of the muscle exhibited a slow component, that was absent if the muscle was curarized and stimulated directly. The observation that direct stimulation per se in a curarized muscle was insufficient to activate the slowly relaxing fibers indicates that these muscle cells are incapable of generating propagated action potentials and that they require the integrity of neuromuscular transmission to develop tension. Therefore, it is possible to conclude that the slow relaxation phase represents the decline of tonic fiber contraction. In these experiments the contribution of tonic fibers to the total tension of the muscle appeared to be rather small, about 10%, but this might be underestimated since the duration of the tetanic stimulation used by Close and Luff was about 200 msec, not long enough for the fibers to reach maximal tension (Matyushkin, 1964; Browne, 1976).

Lennerstrand (1974) recorded the tension developed by four motor units that had been identified with extracellular recording as being composed of multiply innervated fibers without action potentials. He found that the fusion frequency was below 100 per sec and the maximal tetanic tension required stimulation rates of 150–200 per sec, which is well above fusion frequency.

A detailed study of slow-contracting fibers in the extraocular muscle of sheep has been published recently by Browne (1976). Using the anodal block technique developed by Floyd (1970) to stimulate selectively the high-threshold motor axons, he showed with extracellular recording that stimulation of the trochlear nerve evoked a compound nerve action potential with fast and slow peaks, each having a different threshold. The fast peak with a low threshold was gradually blocked by increasing the intensity of the anodal stimulation, and at 33 times threshold intensity it disappeared so that only the slow-conducting axons remained active. Electromyograms of the superior oblique at different intensities of anodal stimulation revealed the presence of a short-latency biphasic potential and a long-latency monophasic wave. The first appeared at low stimulation intensities and was blocked by high-intensity stimuli, indicating that the muscle fibers innervated by the large-diameter axons could generate propagated action potentials. The late monophasic potential appeared at high intensities and persisted after complete block of the biphasic signal. Thus, muscle fibers innervated by small-diameter axons are the site of origin of the monophasic wave. Browne (1976) has indicated that the monophasic nature of this signal cannot be taken as evidence that these fibers have only small junction potentials since in birds monophasic potentials can be recorded from multiply innervated fibers that give rise to propagated action potentials when indirectly stimulated (Ginsborg, 1960). In Browne's study (1976), anodal block stimulation of the muscle with a single shock intense enough to block the fast-conducting axons, induced a muscle tension that was small in amplitude and slow in time course. This contractile response had contraction and half-decline times of about 40 and 75 msec, respectively. The magnitude of this slow tension was only 3% of the twitch tension evoked by conventional (i.e., cathodal), supramaximal stimulation of the whole muscle. The comparison of the time course of the twitch and slow tension clearly demonstrates that the slowly contracting fibers, while not influencing significantly the peak of the twitch may contribute to its falling phase, giving rise to a slow component during twitch relaxation. Matyushkin (1964) observed in rabbit eye muscles a similar slow phase in the relaxation of the twitch.

Browne (1976) showed that anodal block stimulation of the small diameter axons at a frequency of 10 per sec was enough to produce partial summation of the slow contractions and apparent fusion was found at a frequency of 60 per sec. Maximal tension was evoked at stimulation rates higher than the fusion frequency, between 120–150 per sec, in agreement with the finding of Lennerstrand (1974). The maximal velocity of tension development was reached at still higher rates: 170 and 200 per sec in two experiments (Browne, 1976). The mean maximal tension developed by these fibers was 8.8 gm, about 5% of the maximal tetanic tension. The resistance to fatigue of the slowly contracting fibers was remarkable, stimulation of all the fibers at a rate of 80 per sec for 16 sec showed a decline of the tetanus of 49% but when the slowly contracting fibers were selectively stimulated at the same rate and for the same period, the tension decreased by only 9%.

Browne (1976) has pointed out that the failure of Barmack *et al.* (1971), and Fuchs and Luschei (1971a) to detect the summation of slow contractions using a stimulation frequency of 30 per sec can be explained by his observation that at this rate of stimulation, a tension amounting only to 10–15% of the whole muscle twitch tension can be recorded. Such a contraction would not have been obvious under the experimental conditions of these authors. The fused tensions recorded by Hess and Pilar (1963), who stimulated selectively the tonic system at only 30 stimuli per sec (Fig. 10), can be explained by a reduction of the rate of rise and decline of the tension brought about by the temperature of 20°–25°C at which these experiments were performed. Close and Hoh (1968) have demonstrated in rat limb muscle that at 25°C the contraction and half-relaxation times are about twice their values at 35°C.

It is not known whether the slow tensions described by Browne (1976) are evoked by propagated action potentials or by small junction potentials. Therefore, it is not possible to classify the activated fibers either as slow-twitch fibers or as tonic fibers. There are, however, reasons that justify their inclusion here among the tonic or MINC fibers. First, the tensions are very slow, with contraction and half-relaxation times approximately twice those of the slow-twitch fibers described previously (Bach-y-

Rita and Ito, 1966; Lennerstrand, 1974). Second, the time to reach maximal tension with repetitive stimulation at 100 per sec is 0.7 sec, identical to the value reported by Matyushkin (1964) for rabbit tonic fibers under the same regime of stimulation.

The literature on the physiological properties of eye-muscle fibers reviewed in the preceding sections indicates that muscle fibers generate two types of electrical signals that evoke two kinds of contractile responses. These are the twitch and the tonic tensions and, as was discussed, they differ in the rates at which the fibers contract and relax. It should be stressed that this difference between the two types of fibers is not only the consequence of their electrophysiological properties but is also the consequence of their different intrinsic properties that are largely unknown and may involve, among other phenomena, variations in excitation–contraction coupling, in the molecular interaction of the contractile proteins, etc. It is known, however, that such properties are largely controlled by the trophic influence that the central nervous system exerts on muscle (Guth, 1968; Harris, 1974; Nelson, 1975).

The remarkable differences in the contractility of the two systems is fully disclosed when the fibers are depolarized in a sustained manner. Chiarandini (1976) studied this aspect *in vitro,* exposing the rat inferior rectus muscle to salines with elevated K^+ concentrations. He observed that 20 m*M* K^+ produced threshold tensions that became maximal with 50 m*M* K^+ and that K^+ concentrations between 30 and 50 m*M* produced responses consisting of a rapidly rising transient tension that was followed by a sustained tension. This last tension could have a duration of several hours and showed a small degree of relaxation as long as the muscle was depolarized with elevated K^+ (Fig. 11). In control experiments with dia-

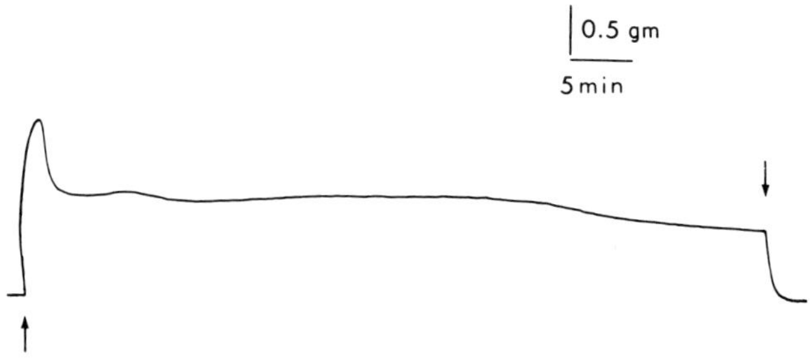

FIG. 11. Tension induced by an elevated K concentration in rat inferior rectus muscle. The exposure of the muscle to a saline containing 50 m*M* K (first arrow, left) produced a rapid tension increase that subsided in a few minutes and was followed by a maintained tension. This tonic response relaxed very slowly until the test saline was replaced by normal saline (second arrow, right). (From Chiarandini, 1976.)

phragm muscle, that has no tonic fibers, K^+ induced only transient tensions. The initial phasic component of the extraocular-muscle responses was attributed to twitch fibers while the long-lasting tension was ascribed to the activation of tonic fibers. These observations coincide with the responses of twitch and tonic fibers of frog to K^+-induced depolarizations (Kuffler and Vaughan Williams, 1953b) and emphasize once more the similarities between these muscle cells and extraocular muscle fibers. Chiarandini (1976) observed that 30 m*M* K^+, which should induce a depolarization of about 45 mV, evoked 80% of the maximal tension in tonic fibers and only 50% in twitch fibers. The fact that in the first group of fibers a moderate depolarization is enough to evoke a large tension is in keeping with the postulation that tonic fibers are activated by summated sjps and not by action potentials. The maximal sustained tension was about 50% of the tetanic tension in the same muscles. In cat eye muscles, Bach-y-Rita and Ito (1966) found that the maximal tension evoked by succinylcholine, which also activates tonic fibers, is about one-third of the tetanic tension. In addition, Chiarandini (1976) demonstrated that the sustained tension was highly dependent upon the Ca^{2+} concentration in the bathing saline which suggests that in tonic fibers most of the Ca^{2+} required to initiate the contraction comes from the extracellular space and not from the SR as in the case of twitch fibers. The dependence upon external Ca^{2+} concentration is in agreement with the finding of very sparse elements of SR and poorly developed transverse tubules in the multiply innervated fibers of rat extraocular muscles (Mayr, 1971; see Section I,B,1 and 2).

3. Possible Role of Fiber Types in Ocular Motility

Eyes are able to perform a variety of movements characterized by widely differing speeds. These range from slow vergence during which eyes reach a maximal velocity of not more than about 10 degrees per sec to the fast saccades that involve a maximal velocity of as much as 400 degrees per sec (Robinson, 1966; 1968). The existence in EOM of fibers with different morphological and contractile characteristics has raised the question as to whether these different muscle fibers may respectively relate to the various types of eye movements. With our present information it remains premature (see Peachey, 1971) to try to correlate the six "types" of fibers summarized in Table I with the electrical and contractile properties previously described. At the level of organized ocular motility, however, there are several lines of evidence concerning the involvement of the different fiber populations.

The suggestion that the slow-contracting tonic muscle fibers subserve vergence while twitch fibers are the basis for the saccadic movements (Alpern and Wolter, 1956; Jampel, 1967) has not been substantiated by electrophysiological studies of single units in oculomotor nuclei since each motorneuron apparently participates in all types of ocular movements. The recruitment and degree of activity of neurons seem to be related to eye position and to its rate of change during a given movement of the eye (Fuchs and Luschei, 1970, 1971b; Robinson, 1970; Schiller, 1970; Keller and Robinson, 1972). This suggests that regardless of their contractile properties, all fibers of eye muscles may participate in all ocular movements.

This question has also been examined by inducing eye movements and recording extracellularly the concomitant electrical activity of the muscle. Scott and Collins (1973) studied the electromyogram during fixation, tracking, vergence, and saccadic movements and found that motor units that changed their activity during a given type of ocular movement also participated in other types of movements or during fixation at another gaze position. The recruitment of the units occurred earlier in the orbital layer and later in the global region of the muscles, thus anticipating Barmack's findings (see Section II,C,1). Scott and Collins (1973) concluded that there is a spectrum of involvement of the motor units. Some units provide a rather steady low-level force for the primary position and have a slow increase in activity with gaze into the field of action. In contrast, other units participate only in the extreme gaze or saccadic movement into the field of action of the muscle. They also concluded that the various types of fibers are functionally differentiated on the basis of the amount of work they do, rather than on the basis of the type of eye movement to which they contribute. It should be noted, however, that the generality of the conclusions of Scott and Collins (1973) is limited by their having recorded only the action potentials or "spikes" of muscle fibers insofar as it has been found that such electrical activity does not occur in tonic fibers (Pilar, 1967; D. J. Chiarandini and E. Stefani, unpublished observations).

The possible role of tonic fibers in ocular movements remains obscure. Their remarkable ability to contract in a sustained manner (Chiarandini, 1976) and their resistance to fatigue during repeated stimulation (Lennerstrand, 1974; Browne, 1976) suggest that they might be involved in the generation of holding tension. Along similar lines, Browne (1976) found that activation of slowly contracting fibers increases the resistance of the superior oblique muscle to small-amplitude stretches leading him to suggest that tonic fibers may smooth and damp the action of antagonistic muscles during fixation. Thus, the possibility remains that the tonic motor system plays a preponderant role in some selective aspects of eye movements.

The recent description in the OSL of rabbit EOM of fibers having systematic morphological variability along their length and of multiply innervated fibers with end platelike junctions (Davidowitz *et al.*, 1977a,b) is evidence of a further degree of EOM structural complexity that may have functional implications. Such unusual structural findings would be compatible with the notion previously advanced by Yellin (1969) and Peachey (1971), that polyneural innervation in EOM might result in the occurrence of both graded local tensions and action potentials in the same fiber. It is furthermore conceivable that such duality of response might be selectively confined to different regions of the same muscle fiber (Davidowitz *et al.*, 1977b). The presence of both superficial and end platelike junctions on some multiply innervated fibers would also make them comparable, in this respect, to some intrafusal fibers (Barker, 1974). Thus, the possibility cannot be excluded that such EOM fibers in rabbit might play a functional role analogous to that of intrafusal fibers. Muscle spindles as such have not been observed in EOM of rabbit but are found in various numbers in some other species (see Barker, 1974).

In the attempt to unravel the considerable morphological and functional complexities of EOM and elucidate the role of different fiber populations, two experimental approaches seem most promising of success, i.e., the simultaneous recording of tension and electrical activity from regions of the muscle defined as precisely as possible in terms of fiber population content, and the correlation of electrical activity and structure of individual fibers by means of intracellular recording and dye marking for subsequent morphological identification.

Acknowledgments

The authors thank the helpful comments of Drs. A. Bondi and C. De Lemos and the valuable assistance of Ms. K. Rhee during the preparation of the manuscript. Work from the authors' laboratory described in this review was supported by National Institutes of Health Grants EY 01297 and EY 00309.

References

Alpern, M., and Wolter, J. R. (1956). *Arch. Ophthalmol.* **56**, 685–690.

Alvarado, J., and Van Horn, C. (1975). *In* "Basic Mechanisms of Ocular Motility and Their Clinical Implications" (G. Lennerstrand and P. Bach-y-Rita, eds.), pp. 15–45. Pergamon, Oxford.

Bach-y-Rita, P. (1971). *In* "The Control of Eye Movements" (P. Bach-y-Rita, C. C. Collins, and J. E. Hyde, eds.), pp. 7–45. Academic Press, New York.

Bach-y-Rita, P. (1975). *In* "Basic Mechanisms of Ocular Motility and Their Clinical Implications" (G. Lennerstrand and P. Bach-y-Rita, eds.), pp. 91–111. Pergamon, Oxford.

Bach-y-Rita, P., and Ito, F. (1966). *J. Gen. Physiol.* **49**, 1177–1198.

Bach-y-Rita, P., and Lennerstrand, G. (1975). *J. Physiol. (London)* **244**, 613–624.

Barker, D. (1974). *In* "Handbook of Sensory Physiology" (C. C. Hunt, ed.), Vol. III (Muscle Receptors)/2, pp. 1–190. Springer-Verlag, Berlin and New York.

Barmack, N. H. (1977). *J. Neurophysiol.* **40,** 779–790.
Barmack, N. H., Bell, C. C., and Rence, B. G. (1971). *J. Neurophysiol.* **34,** 1072–1079.
Barnard, R. J., Edgerton, V. R., Furukawa, T., and Peter, J. B. (1971). *Am. J. Physiol.* **220,** 410–414.
Boyd, I. A., and Martin, A. R. (1956). *J. Physiol. (London)* **132,** 61–73.
Boyd, I. A., and Martin, A. R. (1959). *J. Physiol. (London)* **147,** 450–457.
Breinin, G. M. (1971). *Am. J. Ophthalmol.* **72,** 1–9.
Brown, G. L., and Harvey, A. M. (1941). *J. Physiol. (London)* **99,** 379–399.
Browne, J. S. (1976). *J. Physiol. (London)* **254,** 535–550.
Burke, R. E. (1967). *J. Physiol. (London)* **193,** 141–160.
Burke, W. (1957). *J. Physiol. (London)* **135,** 511–521.
Burke, W., and Ginsborg, B. L. (1956). *J. Physiol. (London)* **132,** 599–610.
Chiarandini, D. J. (1976). *J. Physiol. (London)* **259,** 199–212.
Close, R. I. (1967). *J. Physiol. (London)* **193,** 45–55.
Close, R. I. (1969). *J. Physiol. (London)* **204,** 331–346.
Close, R. I. (1972). *Physiol. Rev.* **52,** 129–197.
Close, R. I., and Hoh, J. F. Y. (1968). *Nature (London)* **217,** 1179–1180.
Close, R. I., and Luff, A. R. (1974). *J. Physiol. (London)* **236,** 259–270.
Collins, C. C. (1971). *In* "The Control of Eye Movements" (P. Bach-y-Rita, C. C. Collins, and J. E. Hyde, eds.), pp. 283–325. Academic Press, New York.
Cooper, S., and Eccles, J. C. (1930). *J. Physiol. (London)* **69,** 377–385.
Cooper, S., and Fillenz, M. (1955). *J. Physiol. (London)* **127,** 400–413.
Costantin, L. L. (1968). *J. Physiol. (London)* **195,** 119–132.
Davidowitz, J., Philips, G., Pachter, B. R., and Breinin, G. M. (1975a). *Am. J. Pathol.* **78,** 191–196.
Davidowitz, J., Philips, G., Pachter, B. R., and Breinin, G. M. (1975b). *J. Ultrastruct. Res.* **51,** 307–313.
Davidowitz, J., Pachter, B. R., and Breinin, G. M. (1976a). *34th Annu. Proc. Electron Microsc. Soc. Am., Miami Beach, Florida*, pp. 110–111.
Davidowitz, J., Pachter, B. R., and Breinin, G. M. (1976b). *Stain Technol.* **51,** 139–140.
Davidowitz, J., Pachter, B. R., and Breinin, G. M. (1976c). *Stain Technol.* **51,** 247–248.
Davidowitz, J., Pachter, B. R., Philips, G., and Breinin, G. M. (1977a). *35th Annu. Proc. Electron Microsc. Soc. Am., Boston, Mass.*, pp. 566–567.
Davidowitz, J., Philips, G., and Breinin, G. M. (1977b). *Invest. Ophthalmol.* **16,** 711–729.
Del Castillo, J., and Katz, B. (1954). *J. Physiol. (London)* **124,** 560–573.
Duke-Elder, W. S., and Duke-Elder, P. M. (1931). *Proc. R. Soc. London, Ser. #B* **107,** 332–343.
Fatt, P., and Katz, B. (1951). *J. Physiol. (London)* **115,** 320–370.
Floyd, K. (1970). *J. Physiol. (London)* **208,** 47–48P.
Fuchs, A. F., and Luschei, E. S. (1970). *J. Neurophysiol.* **33,** 382–392.
Fuchs, A. F., and Luschei, E. S. (1971a). *J. Physiol. (London)* **219,** 155–166.
Fuchs, A. F., and Luschei, E. S. (1971b). *Exp. Brain Res.* **13,** 78–89.
Ginsborg, B. L. (1960). *J. Physiol. (London)* **154,** 581–598.
Goldberg, S. J., Lennerstrand, G., and Hull, C. D. (1976). *Acta Physiol. Scand.* **96,** 58–63.
Guth, L. (1968). *Physiol. Rev.* **48,** 645–687.
Hanson, J., and Lennerstrand, G. (1977). *Acta Ophthalmol.* **55,** 88–102.
Harker, D. W. (1972). *Invest. Ophthalmol.* **11,** 956–969.
Harris, A. J. (1974). *Annu. Rev. Physiol.* **36,** 251–305.
Hess, A. (1961). *J. Cell Comp. Physiol.* **58,** 63–80.
Hess, A., and Pilar, G. (1963). *J. Physiol. (London)* **169,** 780–798.

Hess, A. (1974). *In* "The Peripheral Nervous System" (J. I. Hubbard, ed.), pp. 75–86. Plenum, New York.
Hines, M. (1931). *Am. J. Anat.* **47,** 1–54.
Hodgkin, A. L., and Rushton, W. A. H. (1946). *Proc. R. Soc. London, Ser. B* **133,** 444–479.
Huerta, M., and Stefani, E. (1975). *Actas 18th Congr. Nac. Cienc. Fisiol., San Luis de Potosi, Mexico*, p. 79.
Jampel, R. S. (1967). *Invest. Ophthalmol.* **6,** 288–293.
Kato, T. (1938). *Okajimas Folia Anat. Jpn.* **16,** 131–145.
Keller, E. L., and Robinson, D. A. (1972). *Vision Res.* **12,** 369–382.
Kiyohara, T., and Sato, M. (1967). *Jpn. J. Physiol.* **17,** 720–725.
Kuffler, S. W., and Gerard, R. W. (1947). *J. Neurophysiol.* **10,** 383–394.
Kuffler, S. W., and Vaughan Williams, E. M. (1953a). *J. Physiol. (London)* **121,** 289–317.
Kuffler, S. W., and Vaughan Williams, E. M. (1953b). *J. Physiol. (London)* **121,** 318–340.
Kupfer, C. (1960). *J. Physiol. (London)* **153,** 522–526.
Lennerstrand, G. (1974). *Acta Physiol. Scand.* **91,** 458–474.
Lennerstrand, G. (1975). *In* "Basic Mechanisms of Ocular Motility and Their Clinical Implications" (G. Lennerstrand, and P. Bach-y-Rita, eds.), pp. 119–143. Pergamon, Oxford.
Lennerstrand, G., and Bach-y-Rita, P. (1974). *Invest. Ophthalmol.* **13,** 879–882.
Lennerstrand, G., and Bach-y-Rita, P., eds. (1975). "Basic Mechanisms of Ocular Motility and Their Clinical Implications." Pergamon, Oxford.
Liley, A. W. (1956). *J. Physiol. (London)* **132,** 650–666.
Matyushkin, D. P. (1961). *Sechenov Physiol. J. USSR.* **47,** 65–69.
Matyushkin, D. P. (1964). *Fed. Proc., Fed. Am. Soc. Exp. Biol.* **23,** T1103–1106.
Matyushkin, D. P., and Drabkina, T. M. (1970). *Neurosci. Transl.* **15,** 121–127.
Mayr, R. (1971). *Tissue & Cell* **3,** 433–462.
Mayr, R., Gottschall, J., Gruber, H., and Neuhuber, W. (1975). *Anat. Histol. Embryol.* **148,** 25–34.
Miller, J. E. (1967). *Invest. Ophthalmol.* **6,** 18–39.
Nelson, P. G. (1975). *Physiol. Rev.* **55,** 1–61.
Nemet, P., and Miller, J. E. (1968). *Invest. Ophthalmol.* **7,** 592–598.
Ozawa, T. (1964). *Jpn. J. Ophthalmol.* **8,** 111–116.
Ozawa, T., Cheng-Minoda, K., Davidowitz, J., and Breinin, G. M. (1969). *Doc. Ophthalmol.* **26,** 192–201.
Pachter, B. R., Davidowitz, J., and Breinin, G. M. (1976). *Tissue & Cell* **8,** 547–560.
Peachey, L. (1971). *In* "The Control of Eye Movements" (P. Bach-y-Rita, C. C. Collins, and J. E. Hyde, eds.), pp. 47–66. Academic Press, New York.
Peachey, L. D., Takeichi, M., and Nag, A. C. (1974). *In* "Exploratory Concepts in Muscular Dystrophy" (A. T. Milhorat, ed.), Vol. II, pp. 246–254. Excerpta Medica Foundation, Amsterdam.
Pilar, G. (1967). *J. Gen. Physiol.* **50,** 2289–2300.
Robinson, D. A. (1966). *J. Pediatr. Ophthalmol.* **3,** 31–37.
Robinson, D. A. (1968). *Science* **161,** 1219–1224.
Robinson, D. A. (1970). *J. Neurophysiol.* **33,** 393–404.
Schiller, P. H. (1970). *Exp. Brain Res.* **10,** 347–362.
Schulze, F. (1977). *Albrecht von Graefes Arch. Klin. Exp. Ophthalmol.* **203,** 31–43.
Scott, A. B., and Collins, C. C. (1973). *Arch. Ophthalmol.* **90,** 319–322.
Stefani, E., and Steinbach, A. B. (1969). *J. Physiol. (London)* **203,** 383–401.
Stein, J. M., and Padykula, H. A. (1962). *Am. J. Anat.* **110,** 103–124.
Teräväinen, H. (1968). *Z. Zellforsch. Mikrosk. Anat.* **90,** 372–388.
Yellin, H. (1969). *Exp. Neurol.* **25,** 153–163.
Yellin, H., and Guth, L. (1970). *Exp. Neurol.* **26,** 424–432.

Binding Proteins for Retinol in Retina and Pigment Epithelium

GIOVANNI MARAINI[1]

I. Introduction 143
II. The Transport Mechanism for Vitamin A in Plasma 145
III. The Role of the Pigment Epithelium in the Metabolism of Retinol within the Eye 149
A. Uptake of Retinol from Blood 150
B. Transfer of Retinol to and from the Photoreceptors . . . 151
IV. The Interaction of the Retinal Pigment Epithelium with Circulating Retinol-Binding Protein and the Uptake of Retinol 153
V. Cellular Retinol-Binding Proteins in Extraocular Tissues . . . 158
VI. Binding Proteins for Retinol in Retinal Pigment Epithelium . . 160
A. Membrane Proteins 160
B. Cytosol Proteins 164
VII. Binding Proteins for Retinol in the Retina and Photoreceptor Cell Fractions 167
VIII. Considerations and Summary 169
References 172

I. INTRODUCTION

The metabolism of retinol is a process of primary importance in the eye where the aldehyde of vitamin A (retinal) constitutes the chromophore of visual pigment molecules (Wald, 1933, 1934, 1935, 1937).

The difference among the various visual pigments of vertebrate retinas appears to reside entirely in the protein part of the molecule. Both the aldehyde of vitamin A_1 (retinal) and the aldehyde of vitamin A_2 (3-dehydroretinal) can constitute the prosthetic group of visual pigments.

[1] Department of Ophthalmology, University of Parma, Parma, Italy.

ISBN 0-12-153001-9

Vitamin A_2 visual pigments (porphyropsin and cyanopsin) are characteristics of most aquatic vertebrates while vitamin A_1 visual pigments (rhodopsin and the corresponding cone pigments iodopsins) are present as the protein components of the outer segment membranes of photoreceptors in the retinas of birds and mammals. Both types of visual pigments may be present in the retinas of some amphibians and teleost fishes.

In all known visual pigments in the unbleached state the retinal or 3-dehydroretinal molecule takes the form of the 11-cis isomer (Bridges, 1970); all-trans isomers emerge as the result of light absorption by visual pigments and subsequent photic bleaching. The 11-cis configuration of the chromophore is required again for the regeneration of the pigments from the corresponding opsins (Hubbard and Wald, 1952; Wald *et al.*, 1955).

The utilization of vitamin A within the eye involves therefore a complex set of metabolic steps that include the uptake from the blood, probably the storage as a retinyl ester (Hubbard and Dowling, 1962; Futterman and Andrews, 1964; Berman *et al.*, 1974), the conversion from all-*trans* retinol to 11-*cis* retinal for visual pigment synthesis and the isomerization of all-*trans* retinal to the 11-cis form during pigment regeneration.

Available experimental evidence together with histological and ultrastructural considerations indicate that retinal pigment epithelium must play a key role at least in some of these metabolic steps. In contrast to retinal capillaries which have thick walls with no fenestrations and very few if any pinocytotic vesicles (Cunha-Vaz *et al.*, 1966), choroidal capillaries have thin and fenestrated walls and a high protein permeability which results in a high protein concentration in the choroidal extravascular space (Bill, 1976). Retinal pigment epithelial cells, which are attached to each other by tight junctions (Shiose, 1971), constitute a part of the blood–retinal barrier and probably regulate the flow of tissue fluid to and from the photoreceptor cells.

One important aspect of vitamin A metabolism within the eye must therefore involve the process of the transport of this molecule from the extravascular space of the choriocapillaris to the pigment epithelium and the transfer of retinol (or its derivatives) from these cells to the photoreceptors and back.

In recent years some experimental data have accumulated that have considerably clarified the mechanism by which circulating retinol is transferred from its carrier plasma protein to retinal pigment epithelium. The identification of some proteins able to bind retinol in the soluble as well as in the membrane fractions of these cells seems to be a promising approach to the problem of the transport of the chromophore to and from the photoreceptors, a process which still remains fundamentally unknown.

Independently from the obvious biological interest of this problem, its solution could possibly be of relevant significance for a better understanding of the metabolic defect at the root of a large group of genetic retinal dystrophies resulting in the death of the photoreceptor cell population.

This article will be an attempt to summarize recent work and an effort will be made to stress some points of particular interest in the study of the uptake and transport of retinol in the neuroepithelium of the eye.

II. THE TRANSPORT MECHANISM FOR VITAMIN A IN PLASMA

Vitamin A alcohol (Fig. 1) has a molecular weight of 286, an absorption maximum of 324–325 nm, and its solutions exhibit a characteristic green fluorescence. It is practically insoluble in water, and dissolves easily in absolute alcohol, methanol, chloroform, ether, fats, and oils. Free retinol is a rather unstable compound, especially in an aqueous environment, and is rapidly inactivated by ultraviolet light.

Dietary vitamin A is absorbed as retinyl esters from the intestines via the lymphatic route and transported, mainly in association with the chylomicrons, to the liver which represents the main storage site for the vitamin (Huang and Goodman, 1965; Goodman *et al.*, 1965; Fidge *et al.*, 1968). Linder *et al.* (1971) reported that virtually all vitamin A is stored in the hepatocytes, a result subsequently confirmed by Peterson *et al.* (1973) in rats intravenously injected with ^{3}H-retinyl acetate by studying liver cell separated on colloidal silica-gel gradients. In normal rats the liver already contains its maximal amount of radioactive vitamin A 15 min after the intravenous administration of lipoprotein-bound ^{3}H-retinyl acetate. It is only slowly depleted of the radioactivity since it retains two-thirds after 16 hr (Peterson *et al.*, 1973).

Before the newly absorbed vitamin A can appear in the circulation a number of metabolic processes must occur; these include hydrolysis of retinyl ester to provide retinol and the formation of a complex between retinol and its plasma transport protein.

This process probably does not involve a *de novo* synthesis of the transport protein, but more likely involves the release into the circulation

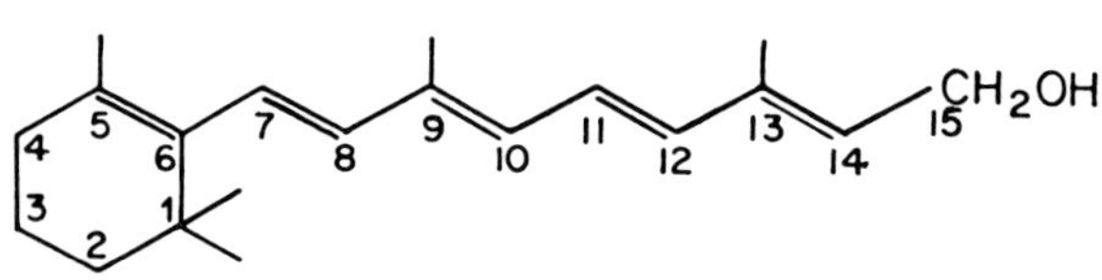

FIG. 1. Retinol.

TABLE I

The Amino Acid Composition of RBP

Amino acid	Residues per RBP molecule	
	Man[a]	Man[b]
Lysine	10–11	10
Histidine	2	2
Arginine	14–15	14
Aspartic acid	29	26
Threonine	9–10	9
Serine	11	11
Glutamic acid	21	18
Proline	5–6	6
Glycine	12	11
Alanine	14–15	13
Half-cystine	5–6	5
Valine	14	12
Methionine	3–4	4
Isoleucine	4	4
Leucine	14	12
Tyrosine	8	8
Phenylalanine	10–11	10
Tryptophan	4	6
	185–193	181

[a] From Kanai *et al.* (1968).
[b] From Peterson (1971a).

of preformed carrier from a preexisting pool in the liver; this hypothesis seems to be supported by the short time interval required (45 min) and by the observation that treatment of rats with puromycin or cycloheximide 1 hr before vitamin A injection does not prevent the rise of serum levels of carrier-bound retinol (Smith *et al.*, 1973).

Vitamin A alcohol is therefore present in plasma bound to a specific transport protein called retinol-binding protein (RBP)[2] (Kanai *et al.*, 1968; Peterson, 1971a). This protein has a hydration density greater than 1.21 and consists of a single polypeptide chain with a molecular weight, as estimated by different methods, of approximately 21,000–22,000. The results of amino acid analysis on purified samples of RBP, as reported from different laboratories, are summarized in Table I and indicate an unusually high content of aromatic amino acids.

[2] The abbreviation RBP will be used in the text only to indicate the plasma retinol-binding protein.

The absorption spectrum of native RBP isolated from human serum is shown in Fig. 2. The spectrum demonstrates two peaks of nearly equal height at 280 and 330 nm. The 330 nm peak represents the absorbance of protein-bound retinol and disappears after the removal of retinol from RBP with heptane. Evidence has been presented by several investigators (Kanai *et al.*, 1968; Peterson, 1971a; Vahlquist *et al.*, 1971) that one molecule of RBP binds one molecule of retinol which is the only molecular form of vitamin A physiologically transported by this specific carrier.

Upon excitation of RBP at 280 and 330 nm two types of fluorescence can be obtained: one with a maximum at 335 nm due to the protein intrinsic chromophores, and another with a maximum at 465–470 nm due to retinol (Fig. 3) (Peterson and Rask, 1971; Goodman and Leslie, 1972). The relative intensity of fluorescence of RBP-bound retinol is increased by a factor of 5–15 with respect to that of retinol alone in solution and shifted approximately 15 nm toward shorter wavelengths. These changes have been interpreted as evidence that RBP-bound retinol is relatively fixed and immobilized compared to retinol in solution; this interaction is probably of significant physiological importance not only to solubilize the water-insoluble retinol molecule but also to prevent its chemical degradation.

The concentration of RBP in normal human serum is of the order of 3–4.5 mg/100 ml (Kanai *et al.*, 1968; Peterson, 1971a; Maraini *et al.*, 1975).

Comparative studies (Muto *et al.*, 1973) have shown that mammals in general have a plasma–retinol transport system similar to that of man, though immunologically important differences of RBP do occur among mammalian orders.

In human plasma, RBP circulates as a complex with another larger protein, a thyroxine-binding prealbumin (Kanai *et al.*, 1968); this is a tetrameric protein with a molecular weight of approximately 64,000 (Rask *et al.*, 1971a). The formation of the retinol–protein–protein complex

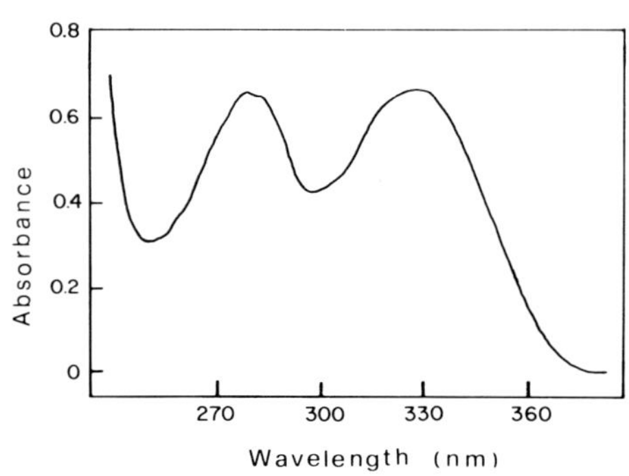

FIG. 2. Absorption spectrum of human RBP.

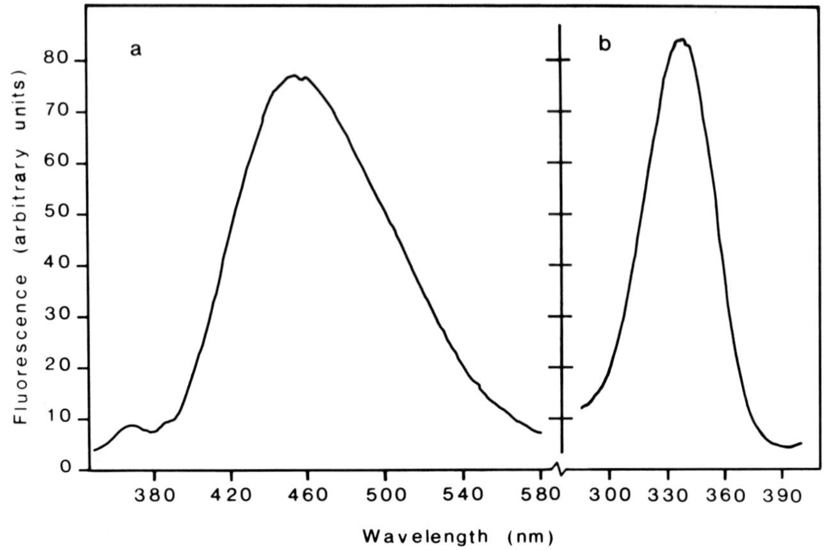

FIG. 3. Fluorescence spectrum of human RBP dissolved in 0.002 *M* Tris-HCl buffer, pH 8.0. (a) Emission spectrum recorded at an excitation wavelength of 330 nm. (b) Excitation spectrum recorded at a light emission of 470 nm.

(molecular weight 85,000) resulting from the interaction of prealbumin with RBP is probably of remarkable importance in preventing the filtration of the low-molecular-weight RBP through the vessel walls in renal glomeruli.

The complex between RBP and prealbumin is dependent on the ionic strength of the solution and the two proteins separate at low salt concentrations; on the basis of studies of the optical rotatory dispersion and circular dichroism, Rask *et al.* (1972) concluded that the dissociation of the prealbumin–RBP complex possibly is due primarily to conformational changes occurring only in the RBP molecular architecture.

The binding of RBP to prealbumin is noncovalent and highly specific; experimental evidence has been presented which suggests that this binding may be influenced by reductive alkylation of disulfide bonds or iodination of the two proteins (Raz *et al.*, 1970), by changes in the functional group of the retinyl moiety (Heller and Horwitz, 1973) and that the prealbumin–RBP interaction probably involves one of the four tryptophan residues of RBP (Horwitz and Heller, 1974). Binding of RBP with prealbumin does not occur in bovine serum (Heller, 1975a,b; Futterman *et al.*, 1976).

According to Peterson (1971b) and to Rask *et al.* (1971b) small amounts of free RBP can be isolated from human serum notwithstanding a twofold molar excess of prealbumin over RBP. This unbound RBP has a lower retinol content and lacks the COOH-terminal arginine, an event that has been interpreted as possibly causing a conformational change in the protein molecule which should account for its low affinity for prealbumin. The same authors suggested that this retinol-free RBP could represent a transitory catabolic form of the carrier which, after having accomplished its physiological role, is on its way to be eliminated or degraded by the kidney. It should be noted, however, that White *et al.* (1972) found lysine to be the C-terminal amino acid of native human RBP; to explain the difference between their data and the results of Peterson (1971a) that indicated arginine as the COOH-terminal amino acid, these authors suggested that a peptide might eventually be lost artifactually during the isolation procedure involving complicated and prolonged purification steps.

Although the retinol isolated from native human RBP is the all-trans isomer, apo-RBP has been shown by Heller and Horwitz (1973, 1974) to form stable 1:1 molar complexes with 9-, 11-, and 13-*cis* retinol, retinals, retinoic acid, retinyl acetate, and retinyl oxime. However, only the retinol isomers and the retinoic acid–RBP complexes were able to bind prealbumin at physiologic ionic strength. Conformational changes of the protein molecule in the other chromophores–RBP complexes as well as in apo-RBP possibly accounted for their inability to complex with prealbumin.

III. THE ROLE OF THE PIGMENT EPITHELIUM IN THE METABOLISM OF RETINOL WITHIN THE EYE

Retinal pigment epithelium is composed of a single layer of pigmented cells that coat the outer surface of the retina and are held firmly together by tight junctions. The retinal pigment epithelium faces with its basal surface the Bruch's membrane and the choriocapillary blood stream, and with its apical microvilli envelopes the tips of the photoreceptor's outer segments. The strategic position of this monolayer of cells explains the key role played by retinal pigment epithelium in the exchange of materials between the blood and the outer retina and vice versa (Miller and Steinberg, 1976; Zadunaisky and Degnan, 1976) and in the regulation and control of visual cell metabolism. Retinal pigment epithelium has also been said to secrete at least part of the mucopolysaccharide matrix separating the photoreceptor's outer segments (Berman, 1964) and its role in the destruction of the outer segment's tips in the normal renewal process of visual cells has now been thoroughly documented (Bairati and Orzalesi, 1963; Young, 1967; Young and Bok, 1969).

For the purpose of our discussion, concerning the role of retinal pigment epithelium in the metabolism of vitamin A, we have conveniently divided the subject into two areas.

A. Uptake of Retinol from Blood

Autoradiographic studies (Young and Bok, 1970) of frogs intravenously injected with ^{3}H-vitamin A or ^{3}H-vitamin A acetate have demonstrated that a diffuse labeling is observed over the pigment epithelium 5 min after the injection. The labeling that appears to be mainly concentrated in the oil droplets of these cells continues through 1 hr and then drops off significantly between 1 hr and 10 days. The same type of study in the rat (Sidman and Dowling, 1963; Sherman, 1970) after oral or intraperitoneal administration of radioactive retinol or retinyl acetate confirms the high concentration of radioactivity in the pigment epithelium.

There is therefore little doubt that the utilization of vitamin A by the neuroepithelium, even of eyes with a vascularized retina, occurs mainly if not exclusively via the pigment epithelium. It has already been stressed that the high protein content of the choriocapillaris extravascular fluid renders the basal surface of pigment epithelial cells easily accessible to circulating RBP molecules.

Retinal pigment epithelium has a high retinol content. A dark-adapted *Rana pipiens* eye contains about 2 μg retinyl esters in its pigment epithelium (Hubbard and Dowling, 1962). In studies carried out on preparations of bovine pigment epithelium free of contamination by outer segments or choroidal remnants the content of retinol plus retinyl esters was 1.9 μg per eye or 481 μg per gm of wet tissue (Berman *et al.*, 1974).

The concentration of retinol in pigment epithelium is then about 70 times greater than in the retina and four to five times higher than in the liver, thus supporting the view that retinal pigment epithelium represents indeed the most important storage depot of vitamin A for the eye. In the dark-adapted frog eye about 1.2–2.3 mol of vitamin A (all-trans and 11-cis isomers of retinol and retinyl esters) is recovered for each mol of rhodopsin contained in the rod outer segments; 96% of vitamin A is recovered from the pigment epithelium, the remaining 4% being roughly equally distributed between the visual cell's outer segments and the remainder of the retina (Bridges, 1976a).

The vitamin A in the oil droplets of retinal pigment epithelium is 100% esterified (Hubbard and Dowling, 1962; Bridges, 1975, 1976a). The principal fatty acids of bovine retinal pigment epithelium retinyl esters have been tentatively identified by Berman *et al.* (1974) as 16:0, 17:0 (or 16:2), 18:0, and 18:2.

Considerable evidence has accumulated that indicates that, at least in some animals, pigment epithelium does isomerize vitamin A since a consistent percentage of it is in the 11-cis form (Hubbard, 1956; Krinsky, 1958; Hubbard and Colman, 1959). However the relative proportions of 11-*cis* and all-*trans* retinyl esters in the retinal pigment epithelium depend on the state of adaptation of the eye. Dowling (1960) was not able to detect 11-*cis* retinol in light-adapted rat pigment epithelium while in Hubbard and Dowling's experiments (1962), even after bleaching of most of the rhodopsin of the frog eye, the amount of the 11-cis ester was still 25%. The 11-cis isomers account for 42–54.5% of the retinyl esters of retinal pigment epithelium of fully dark-adapted frogs and the relative proportions of the 11-cis and all-trans isomers depended on the time spent in the dark after the completion of rhodopsin regeneration (Hubbard and Dowling, 1962; Bridges, 1976a).

B. Transfer of Retinol to and from the Photoreceptors

It was shown by Wald (1935) that most of the retinol produced during the bleaching of rhodopsin in the frog's eye leaves the retina. It is now well documented that a back and forth flow of retinol between the retina and its pigment epithelium occurs during the course of light and dark adaptation, the total amount of vitamin A within the eye (retinol + retinal of rhodopsin + retinyl esters) remaining relatively constant (Hubbard and Colman, 1959).

Most of all-*trans* retinol formed from retinal during the bleaching of rhodopsin rapidly moves from photoreceptor outer segments to pigment epithelium. Both in the rat eye (Dowling, 1960; Noell *et al.*, 1971) and in the frog eye (Bridges, 1976a) as much as 80–87% of this retinol produced during light adaptation is transported to pigment epithelium where it accumulates in the oil droplets in the esterified form. It has been shown by Zimmerman *et al.* (1974) that these changes do occur in the rat eye not only with the use of high intensity bleaching light but also after a single light flash that bleaches approximately 10% of the rhodopsin content.

During dark adaptation the reverse processes essentially occur and the retinol is transported back from the pigment epithelium to the photoreceptors for rhodopsin regeneration (Fig. 4).

Several excellent reviews have discussed this topic extensively (Bridges, 1970; Morton, 1972; Baumann, 1972). The scheme reproduced in Fig. 5 has been recently suggested by Bridges (1976a) as representing the visual cycle in the frog eye and will be utilized as a basis for briefly summarizing what is presently thought to happen in the living eye.

Both bovine (Andrews and Futterman, 1964) and frog (Bridges, 1976a) rod outer segments appear to lack the capacity of synthesizing esters. It is

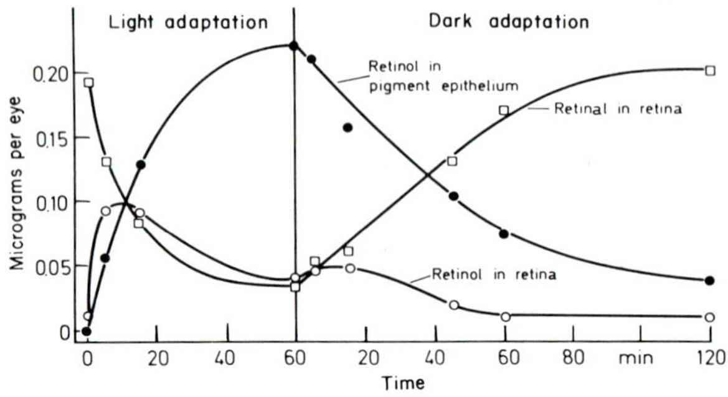

FIG. 4. Distribution of retinol and retinaldehyde in the retina and pigment epithelium of the albino rat eye during light and dark adaptation. (From Dowling, 1960.)

therefore considered that the all-*trans* retinol produced by visual pigment bleaching is actually transported as the alcohol to the retinal pigment epithelium where it will undergo esterification after having entered the cells. The mechanism of this transport is totally unknown. According to Bridges (1976b) rod outer segments are the only structures within the eye where vitamin A exists predominantly as the free alcohol; 60% of this retinol (which on the whole represents only 0.4% of the total retinyl ester) is

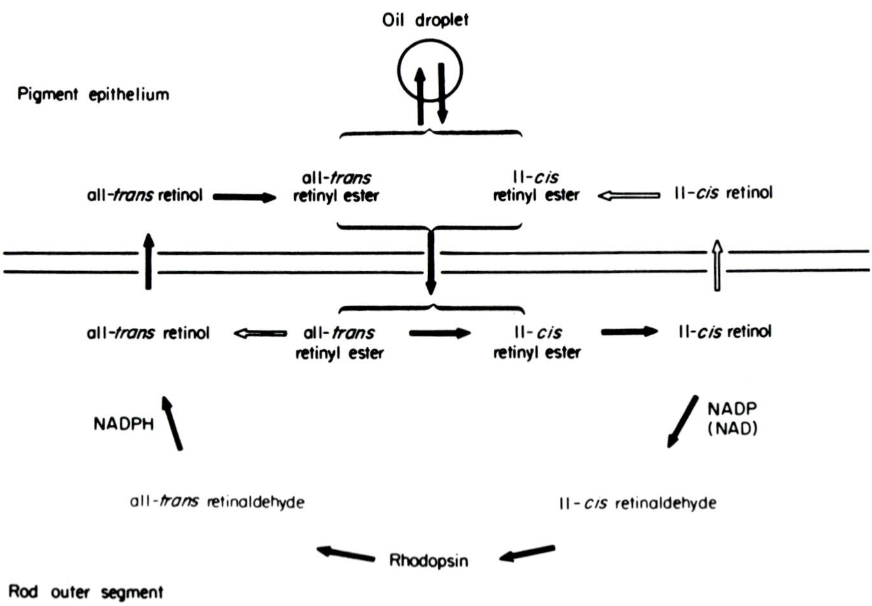

FIG. 5. Suggested "visual cycle" in the frog eye. (From Bridges, 1976a.)

in the 11-cis form and probably constitutes a pool of chromophore groups readily available to recombine with the rhodopsin apoprotein for the production of new visual pigment molecules.

Since under appropriate conditions isolated retina without pigment epithelium (Cone and Brown, 1969) and isolated rod outer segments (Shichi, 1971) can regenerate visual pigment, it has been suggested that for low levels of bleaching as those occurring in the environmental conditions of illumination retinal pigment epithelium might not be essential for rhodopsin regeneration; this would essentially depend on the availability of all-*trans* retinal and on the ability of the photoreceptor cell to isomerize the all-trans to the 11-cis form (Lolley and Schmidt, 1974). It has already been said however that the results of Zimmerman *et al.* (1974) indicate an increase of retinyl esters (which must occur in pigment epithelium) even after bleaching of only 10% of total rhodopsin content thus contrasting with the hypothesis of a substantially different mechanism of rhodopsin regeneration under intense or low illumination conditions.

Sufficient experimental evidence has been presented that in both rat and frog eye (Dowling, 1960; Noell *et al.*, 1971; Bridges, 1976a) the reconstitution of rhodopsin content that takes place during the recovery phase after an intense light adaptation occurs essentially through the utilization of a supply of vitamin A that is represented mainly by the retinyl esters stored in pigment epithelium. According to Bridges (1976a) this flow of vitamin A from the pigment epithelium to the photoreceptors involves, for a complete rhodopsin regeneration, the transport of one-third to one-half of the retinyl esters, mainly in the all-trans form.

It is possible (Baumann, 1972) that vitamin A is actually transferred to the photoreceptors as retinyl esters since isolated retinal pigment epithelium has been found to be ineffective in hydrolyzing retinyl esters. The intimate mechanism of this transfer is however totally unknown.

IV. THE INTERACTION OF THE RETINAL PIGMENT EPITHELIUM WITH CIRCULATING RETINOL-BINDING PROTEIN AND THE UPTAKE OF RETINOL

Some insight into the previously unknown mechanism by which circulating retinol is transferred from its carrier plasma protein to the target cell was obtained independently by Heller (1975a) and by Maraini and Gozzoli (1975) using different types of experimental approach. The conclusion of these investigations has been that an interaction between circulating RBP and some receptor present on the plasma membrane of the target cell is essential for the active uptake of retinol by pigment epithelium.

Both investigations utilized suspension of bovine retinal pigment epithelial cells; these may easily be obtained from freshly enucleated eyes, without the use of proteolytic enzymes and with only minimal mechanical damage, by delicately brushing the inner surface of the posterior eye cup after removal of the retina. When separated under controlled conditions, these cells are reasonably free of contamination and their metabolic integrity is well preserved as indicated by the observation that during short-term incubations (2 hr) they maintain a normal morphology and actively incorporate radioactive amino acids into their proteins.

A more direct line of approach was utilized by Heller (1975a) who investigated the capacity of ^{125}I-labeled human and bovine RBP to interact *in vitro* with isolated bovine pigment epithelium. In an elegant set of experiments he presented convincing experimental evidence that demonstrates that circulating RBP binds specifically to isolated bovine pigment epithelial cells. The binding is rapid and is completed essentially after 1 min, 85% of the binding capacity being already saturated within the first 30 sec. The process is significantly influenced by the incubation temperature and the binding is optimal at 22°C (Fig. 6). Under these experimental conditions an approximate calculation of the second-order binding rate K_1 gave a value of about 10^{-8} M per sec. Specific binding of ^{125}I-RBP was defined by Heller as that amount of radioactive RBP that was displaced from the pigment epithelium cells by the addition of an excess of unlabeled RBP.

The specificity of the interaction of RBP with pigment epithelium was demonstrated by the fact that the presence in the incubation solution of a 600-fold molar excess of different proteins (bovine hemoglobin, bovine serum albumin, horse-liver alcohol dehydrogenase, horse myoglobin, cytochrome c, and human prealbumin) did not influence the binding.

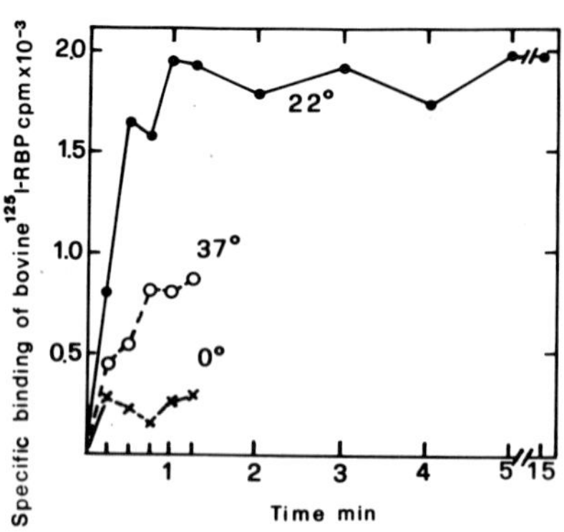

FIG. 6. Specific binding of bovine ^{125}I-RBP to isolated bovine pigment epithelium as a function of temperature. (From Heller, 1975a.)

Both human and bovine RBP bind equally well to bovine pigment epithelium notwithstanding a certain degree of difference between the two proteins, and the observation (Heller, 1975b) that bovine RBP is not complexed with prealbumin in serum. The conclusion that the presence or absence of prealbumin is indifferent to the binding process is confirmed by the same efficacy of specific binding to pigment epithelium showed by ^{125}I-RBP and by ^{125}I-RBP prealbumin complex. This evidence, together with the observation that ^{125}I-RBP which had lost its ability to complex with prealbumin still retained unaffected its capacity to bind to pigment epithelium, has led Heller (1975a) to suggest that the sites of the RBP molecule involved in these two types of interaction are probably different.

In contrast, the presence of retinol in the RBP molecule appears to increase the affinity of the carrier molecule for its pigment epithelium receptor. This seems to be indicated by the decreased effectiveness of apo-RBP, with respect to native RBP, to selectively displace ^{125}I-RBP from bovine pigment epithelium. The suggestion that the reduced activity is in effect due to the absence of retinol in the carrier molecule was based on the normal effectiveness of reconstituted retinol-RBP obtained from the same apo-RBP preparation (Fig. 7).

The possibility of selectively and competitively displacing bound radioactive RBP molecules from pigment epithelium by adding an excess of unlabeled carrier indicates that the interaction of RBP with the target cell is a reversible surface phenomenon and suggests the existence of specific receptors for the plasma RBP on the cell surface.

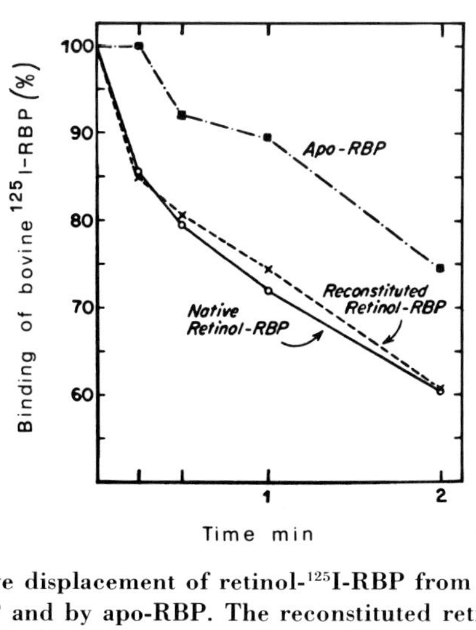

FIG. 7. Competitive displacement of retinol-^{125}I-RBP from pigment epithelium by unlabeled retinol-RBP and by apo-RBP. The reconstituted retinol-RBP was prepared from the apo-RBP utilized in the same experiment. (From Heller, 1975a.)

The specific binding of RBP to pigment epithelium is a linear function of the number of cells present in the incubation solution. From the results of saturation studies the number of retinol-RBP molecules that can bind to a single pigment epithelium cell (and therefore the number of RBP receptor sites on the cell surface) was estimated by Heller as ranging between 3.7×10^4 and 5.2×10^4.

The process of retinol utilization has been investigated by Maraini and Gozzoli (1975) by incubating isolated bovine or human pigment epithelium with human RBP in which ^{3}H-retinol had been substituted for the original chromophore. The ^{3}H-retinol-RBP had absorption and fluorescence characteristics identical to those of the native RBP and retained its capacity to bind prealbumin. Experimental evidence was obtained that pigment epithelium actively binds ^{3}H-retinol only when this molecule is present in the incubation solution bound to its plasma-protein carrier (Table II). The amount of radioactive retinol recovered from pigment epithelium cells and transferred to intracellular water is increased by a factor of approximately 15 when compared to its concentration in the medium. This is suggestive of a process of active transport into the cells against a concentration gradient for which an interaction between circulating RBP and some components of the cell membrane seems to be necessary. Pigment epithelium is not able to utilize free retinol (i.e., not bound to RBP) and under these experimental conditions the concentration of radioactivity in the cells after 2 hr of incubation was reported to be only 0.3–0.5 of that in the external medium.

In good agreement with the kinetic of the process reported by Heller (1975a), binding of carrier bound ^{3}H-retinol to pigment epithelium is very rapid; approximately 60% of the radioactivity present after 2 hr being already recovered in the cells after 2 min. No determinations were made at

TABLE II

***In Vitro* Binding of ^{3}H-Retinol by Pigment Epithelium**[a]

Form of ^{3}H-retinol in the incubation medium	Conc. of ^{3}H-retinol in the medium before incubation (nmol/ml H_2O)	Conc. of ^{3}H-retinol in PE after 2 min (nmol/ml intracellular H_2O)	Conc. of ^{3}H-retinol in PE after 2 hr (nmol/ml intracellular H_2O)	Source of PE
^{3}H-retinol-RBP	0.154	1.637	2.669	Human
^{3}H-retinol-RBP	0.394	2.973	5.403	Bovine
^{3}H-retinol	0.065	0.017	0.036	Bovine
^{3}H-retinol	0.109	0.016	0.033	Bovine

[a] From Maraini and Gozzoli (1975).

shorter time intervals. Under the same experimental conditions, binding of human RBP carried ^{3}H-retinol to human and bovine pigment epithelium is practically identical.

Autoradiographic studies (Heller and Bok, 1976a; Bok and Heller, 1976) confirmed that the interaction between RBP and the retinal pigment epithelium during the process of retinol transport into these cells is in effect a phenomenon occurring on the cell surface. By incubating isolated bovine pigment epithelium and by intravenously injecting ^{125}I-RBP into albino rats, these authors could demonstrate that the radioactivity is restricted to the surface of the cells. Most interestingly the binding of ^{125}I-RBP is limited to the basal choroidal face of the cells while the apical retinal surface does not appear to be labeled. The fact that this nonrandom distribution of radioactivity occurs not only during *in vivo* but also during *in vitro* experiments on isolated cells indicates that an uneven availability of RBP molecules, whose diffusion is hindered by the beltlike tight junctions between the cells, should not be responsible for this phenomenon of uneven distribution of radioactive labeling. A more likely explanation seems to be that the apical surface of pigment epithelium cells lack the RBP receptors that should then be restricted to that basal surface of the cell in contact with the choroidal extravascular space. This does not seem to be the case for pigmented ciliary epithelium that shows a diffuse labeling of all its surfaces (Bok and Heller, 1976).

The absence of radioactivity over the retinal surface of pigment epithelial cells 1 hr after the intravenous injection of labeled RBP was interpreted as evidence of the insignificance of translational movements of the receptor–RBP complex past the region of the zonula occludens and as an indication that RBP cannot be the carrier molecule of retinol to the photoreceptors. This conclusion seems to be supported by the incapacity of photoreceptors to bind radioactive RBP (Heller and Bok, 1976a). This of course does not rule out the possibility that the receptor for RBP on the cell membrane could act at the same time as a receptor for retinol and be involved in some way as a further carrier of the chromophoric molecule.

It seems then to be fairly well established that, in a similar fashion to that demonstrated for several other types of molecules (such as hormones, neurotransmitters, and proteins), circulating RBP has, on the plasma membranes of retinal pigment epithelium, receptor sites with a high affinity and specificity for its substrate. During the interaction between RBP and its receptor, retinol is released from its carrier to some other component of the cell membrane (perhaps the receptor itself) and is taken up by the cell. RBP does not appear to enter the cell and seems to be removed from its binding place by competitive displacement with other RBP molecules. The different binding constants that retinol-RBP and apo-RBP have for the membrane receptor suggest that the overall process of retinol release to the pigment epithelium might be controlled by the different proportions of the two forms of carrier molecules in the extracellular environment and on the cell membrane (Heller, 1975a).

This better knowledge of the intimate process by which circulating retinol is taken up by its target cell in the eye has stimulated similar investigations in those ocular diseases that could possibly be attributed to a derangement in the retinol supply to the retina. The effect of vitamin A deficiency on the eye is known to cause structural deterioration of the outer segments (Dowling and Wald, 1958), and important genetically determined retinal dystrophies in man could *a priori* be associated with a defect of circulating RBP, or with the absence or altered specificity of its receptor on the pigment epithelium.

However, all available evidence so far is negative as to the possibility that such a mechanism might be responsible for the most common form of retinal degeneration in man (recessive retinitis pigmentosa). In this disease neither alteration of circulating RBP in its retinol binding site nor change in its recognition and binding capacity for the receptor of normal pigment epithelium (Fig. 8) could be demonstrated (Maraini, 1974; Maraini and Gozzoli 1976a,b). Of course another possible cause of an impaired retinol supply still has to be ruled out; that is, the absence or altered characteristics of the receptor for RBP on the plasma membranes of pigment epithelial cells.

V. CELLULAR RETINOL-BINDING PROTEINS IN EXTRAOCULAR TISSUES

In 1973 Bashor *et al.* reported the isolation from the high-speed supernatant fraction of rat tissues homogenates of a component able to bind ^{3}H-retinol *in vitro*; this retinol binding substance, present in liver, kidney, testis, and intestinal mucosa, had an estimated molecular weight of 16,000 thereby differing from that reported for rat serum RBP (20,000 according to Muto and Goodman, 1972). Upon incubation of rat tissue cytosols with ^{3}H-retinol the binding was complete within 4 hr and was virtually eliminated by a 200-fold excess of unlabeled retinol; the presence of a similar excess of retinoic acid did not alter the binding which also was only slightly reduced by the same excess of retinal thus indicating a certain degree of binding specificity. The binding component was tentatively identified as a protein on the basis of the observation that treatment of tissue cytosol with Pronase, but not with RNase and DNase, reduced the binding of ^{3}H-retinol to 57% of the controls.

A considerable advance in the study of the soluble retinol-binding pro-

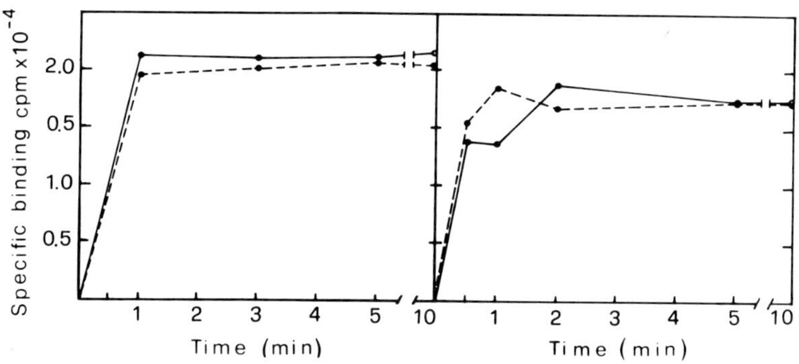

FIG. 8. Specific binding of human ^{125}I-RBP to isolated bovine pigment epithelium (———— RBP from retinitis pigmentosa; -------- RBP from normal subjects). Each experimental point contains 2.5 mg dry weight of cells. (From Maraini and Gozzoli, 1976b.)

tein and of its possible physiological role was achieved in 1974 by Ong and Chytil. By means of chromatographic fractionation studies of rabbit-lung cytosol (the 100,000 *g* supernatant of the tissue homogenate) which had not been previously exposed to any exogenous retinol, these authors reported the separation of three components (A, B, and C) which showed a retinol-type fluorescence. Their molecular weights, as estimated by chromatography on Sephadex G-75, were, respectively, 17,000, 15,000 and 14,000. A partial purification of the B component was achieved and this protein was reported to represent less than 0.002% of the soluble proteins of rabbit lung. Heptane extraction of the protein exibited a fluorescence excitation spectrum identical to that of all-*trans* retinol, and this led Ong and Chytil to conclude that retinol is in fact the naturally occurring ligand of this binding component of rabbit-lung cytosol.

To our knowledge no further purification of these substances has been reported; this would seem to be necessary for a better understanding of their possible physiological role. The difference in molecular weight, together with the reported failure (Bashor and Chytil, 1975) of anti-RBP serum to affect the binding of radioactive retinol to the tissue binding substance, seems however to support the conclusion of a substantial difference between the cytosol and the serum retinol-binding proteins. In the discussion of a subsequent paper (Ong and Chytil, 1975) the same investigators report an endogenous retinol-like fluorescence (excitation maximum at 350 nm, emission maximum at 480 nm) also for native, partially purified, rat-cytosol, retinol-binding protein and "preliminary experiments indicating all-*trans* retinol as the preferred ligand for this protein."

The distribution of the cytosol retinol-binding protein seems to be rather wide; in the rat, the heart and gastrocnemius muscles are the only negative tissues reported so far (Bashor and Chytil, 1975) together with several tumors or tumor cell lines (Novikoff hepatoma, ascites hepatoma AS-30D, mouse Ehrlich ascites tumor, mouse pituitary tumor cell line At T 20).

A smaller protein molecule (of 4800 molecular weight) showing retinol-binding capacity has been isolated from the cytosol fraction of rat-testis seminiferous tubules and partially characterized (Gambhir and Ahluwalia, 1974); its amino acid composition has been reported to be similar to that of blood RBP with the exception of the absence of cysteine and cystine.

Soluble proteins able to bind *in vitro* radioactive retinol with a good degree of specificity have been described also in liver and brain cytosols of rat, guinea pig, cow, and chicken; in some of these tissues more than one retinol-binding protein has been reported, with different sedimentations coefficients (Wiggert and Chader, 1975; Wiggert *et al.*, 1976).

Ong and Chytil (1975) have recently reported the identification, in the 100,000 *g* supernatant fraction of several rat organs incubated with all-*trans*-^{3}H-retinoic acid, of a binding component for retinoic acid. The binding is specific, being nearly inhibited by a 200-fold molar excess of unlabeled vitamin A acid but not by similar additions of retinol, retinal, or oleic acid. That the retinoic acid-binding protein might be a separate protein from the retinol-binding protein of tissue cytosols seems to be supported also by the possibility of partially separating the two binding proteins by chromatography on DEAE-cellulose and by their different tissue distribution.

VI. BINDING PROTEINS FOR RETINOL IN RETINAL PIGMENT EPITHELIUM

A. Membrane Proteins

Maraini *et al.* (1977) have recently reported the identification of a membrane protein able to bind retinol in pigment epithelium. Bovine pigment epithelial cells were incubated for 1 hr in a medium containing ^{3}H-retinol-RBP, washed, homogenized, and fractionated on a discontinuous sucrose gradient. When the gradient was scanned at 280 nm and tested in a scintillation counter, most of the radioactivity was recovered in the two most superficial gradient bands (Fig. 9); these were shown, by electron microscopic examination of the material sedimenting at the different gradient interfaces, to represent the soluble cell sap and the plasma membranes of the cells. The greater amount of radioactivity (as much as 75% of the total) was found in this membrane fraction.

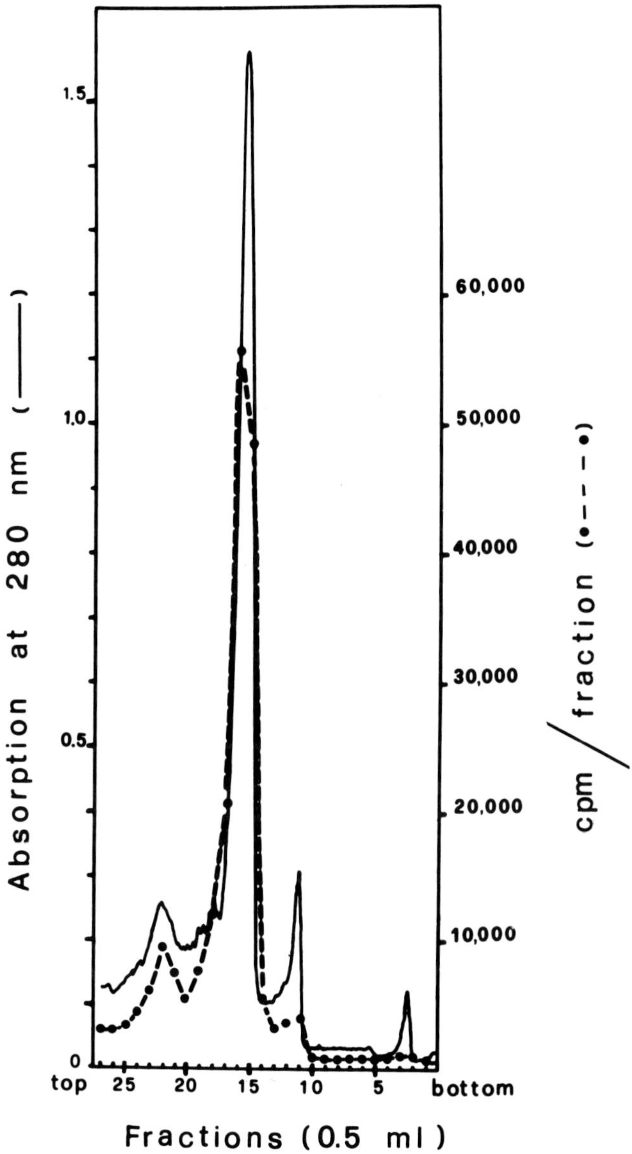

FIG. 9. Scanning and labeling profiles of homogenate of bovine pigment epithelium preincubated with ^{3}H-retinol-RBP and fractionated on a discontinuous sucrose gradient. (From Maraini *et al.*, 1977.)

Collection of the membranes by centrifugation, their solubilization in a detergent solution and their analysis with polyacrylamide gel electrophoresis for protein and radioactivity distribution (Fig. 10) showed that practically all the tritium recovered from the gel corresponded to a single Coomassie blue stained band. The electrophoretic mobility of this protein was reported as being higher than that of human RBP and indicating a molecular weight of approximately 14,500 (Fig. 11). The method employed for the solubilization of the membranes appeared to be crucial for a good recovery in the gel of the membrane-bound radioactivity and very mild conditions were found to be the most effective.

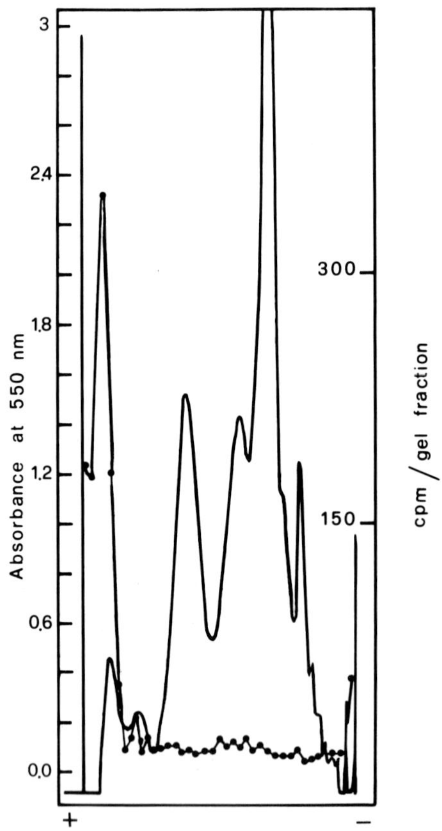

FIG. 10. Staining and labeling profiles of plasma membrane proteins from bovine pigment epithelium preincubated with ^{3}H-retinol-RBP. Plasma membranes were dissolved with 2.5% SDS and 1% dithiothreitol and analyzed for protein and radioactivity distribution by polyacrylamide SDS gel electrophoresis. (From Maraini *et al.*, 1977.)

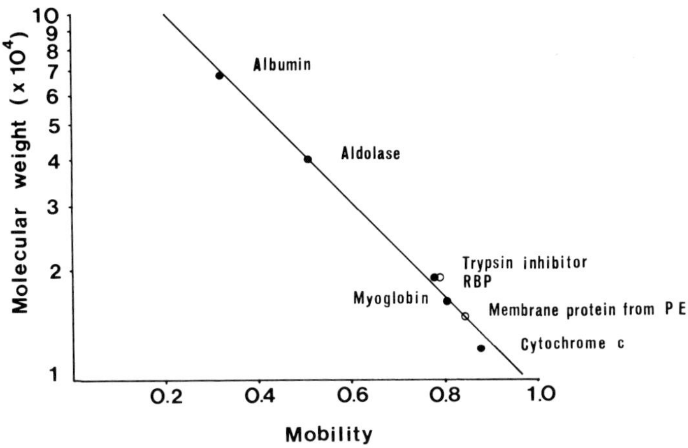

FIG. 11. Determination of the molecular weight of the pigment epithelium membrane retinol-binding protein from a set of six individual standard gels.

Extraction of the radioactive protein band of the gel with lipid solvents and thin-layer chromatography of the concentrated extract showed radioactivity to reside in a spot with mobility characteristics corresponding to those of the reference retinol.

In a parallel set of experiments the authors incubated isolated pigment epithelium in the presence of ^{125}I-RBP, that is, exposed the cells to an RBP solution that contained a labeled carrier protein bound to a nonradioactive chromophore. In these conditions however no appreciable radioactivity was reported as present in the plasma membrane fraction at the end of the isolation procedure.

This observation, together with the lower molecular weight on polyacrylamide gel electrophoresis, induced Maraini *et al.* to suggest that the radioactive protein that they extract from the membranes of cells incubated in the presence of ^{3}H-retinol-RBP could not be the exogenous RBP added to the cell suspension but might represent a membrane protein of pigment epithelium showing retinol-binding properties. The possibility of recovering the same protein band in membrane extracts from native pigment epithelium could support the conclusion that this protein is a normal constituent of these membranes.

Further work is obviously needed before the nature and possible role of this protein can be more precisely defined. It seems however that the experimental evidence so far presented supports the hypothesis that this membrane protein is involved in the process of active retinol transport into pigment epithelium. Nothing can be said presently on the possible relationship of this membrane protein with the RBP receptor present in the membranes of these cells or with the low-molecular-weight, retinol-binding proteins that have been recently demonstrated in the cytosol fraction of retinal pigment epithelium (Wiggert and Chader, 1975).

B. Cytosol Proteins

Retinal pigment epithelium is known to contain large amounts of retinol and its role in the transport of the chromophore (or of a derivative of it) to and from the photoreceptors is well documented. Considering the great lability of the retinol molecule in aqueous solutions and its damaging effect on biological membranes, it was logical to look if also these cells contain, similarly to what has been shown to occur for other tissues of the body, some specific soluble binding protein for retinol with a stabilizing and possibly transport function.

The identification of receptors for retinol in the soluble fraction of retinal pigment epithelium of 11-day-old chick embryos was first reported in 1975 by Wiggert and Chader and confirmed also on guinea pig and cow in a further report from the same laboratory (Wiggert *et al.*, 1976). The studies were carried out *in vitro* by incubating aliquots of tissue cytosol with different concentrations of ^{3}H-retinol and then following the distribution of the absorbance at 280 nm and of the radioactivity after subjecting the cytosol fraction to centrifugation on a sucrose density gradient (5–20%). The distribution of the radioactivity did not follow the 280 nm absorption pattern but was concentrated in two peaks with a sedimentation coefficient of 5.0 S and 2.0 S, respectively. The heavier molecular weight component (5 S) showed a sedimentation constant identical to that of the protein-bound radioactive retinol in serum; brain and liver cytosols also exhibited the same 5 S retinol-binding component and since some plasma contamination during pigment-epithelium separation could not be excluded, it was not possible to decide whether the larger molecular weight binding substance represents a contamination by serum present in the tissue cytosols, or is a separate tissue-binding species or an aggregate of the smaller retinol-binding molecules.

The 2 S receptor seems to have a greater affinity for retinol since its binding capacity is much higher than that of the 5 S component at low retinol concentrations (Fig. 12) and its molecular weight has been reported as ranging from 17,000 to 19,000. The binding capacity for retinol is specific, since a 100-fold excess of unlabeled retinol, but not of retinal, greatly decreases the binding, and reversible, thus indicating its noncovalent nature.

Receptors for retinol with similar sedimentation characteristics have

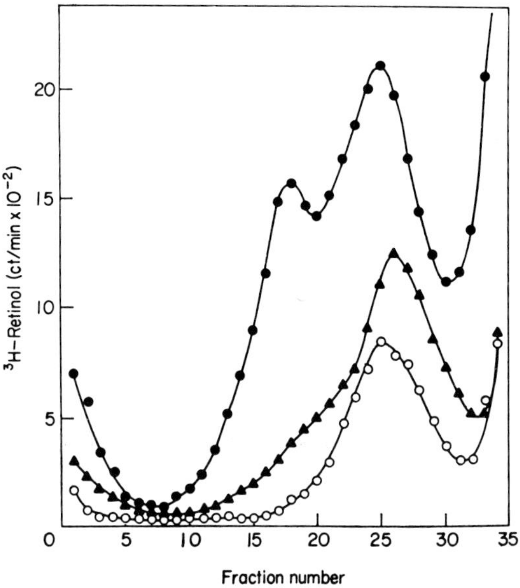

FIG. 12. Binding of ^{3}H-retinol to the 2 S and 5 S receptors of chick pigment epithelium cytosol. Sucrose density gradient centrifugational patterns. The bottom of the gradient tube corresponds to fraction 1 and the top to approximately fraction 35. Concentrations of ^{3}H-retinol in the cytosol were 2×10^{-7} *M* (●——●), 1×10^{-7} *M* (▲——▲), and 5×10^{-8} *M* (○——○). (From Wiggert *et al.*, 1976.)

been demonstrated by the same authors also in the cytosol fractions of brain and liver but of the different tissues studied retinal pigment epithelium has been reported to have the highest concentration of the binding substance.

The available evidence of its protein nature seems to be only indirect and based on the observation (Wiggert and Chader, 1975) that the binding of ^{3}H-retinol to the analogous 2 S retinol receptor of chick-retina cytosol is substantially decreased by Pronase and not by RNase and *N*-ethylmaleimide. No further purification studies have been reported so far.

A similar experimental approach to the problem of the existence of a soluble intracellular binding protein for retinol was utilized by Heller (1976) and by Heller and Bok (1976b) but with results that differ from those obtained by Wiggert and Chader. After *in vitro* incubation of pigment-epithelium cytosol with ^{3}H-retinol and fractionation by gel filtration, Heller (1976) reported that the bound retinol was associated with a protein peak of high moleculear weight with an elution position different from that of plasma RBP and of the RBP–prealbumin complex. Unlike the RBP–prealbumin complex, no dissociation of this high-molecular-weight cytosol–retinol complex occurred at low salt concentrations.

By gel filtration and light-scattering studies, Heller concluded that the molecular weight of this cytosol protein was at least 1.5×10^6 with a calculated diameter of 550 ± 43 Å. Its low density on sucrose-gradient centrifugation suggested that this molecule could be a lipoprotein and in effect it was reported to contain 28–30% (by weight) of extractable lipids. Only cholesterol was contained in its neutral lipid fraction and, in about equal amounts, phosphatidyl choline was found as the only phospholipid. The protein has also been reported to contain glucosamine and galactosamine as well as small amounts of, as yet, undetermined neutral sugars.

The retinol-binding lipoglycoprotein, the amino acid composition of which is reported in Table III, resolved by SDS-polyacrylamide disc electrophoresis in at least two subunits with a molecular weight, respectively, of 35,000 and 150,000. The absorption spectrum of purified samples of the protein, as reported by Heller and Bok, is shown in Fig. 13 and demonstrates a small peak at about 420 nm and a second distinct one at about 280 nm.

TABLE III

Amino Acid Composition of Pigment Epithelial and Rod Outer Segment Cytosol Retinol-Binding Lipoproteins[a]

Amino acid	Pigment epithelium cytosol protein (mol %)	Rod outer segment cytosol protein (mol %)
Lysine	6.64	6.52
Histidine	2.16	2.15
Arginine	4.88	4.69
Aspartic acid	10.65	10.74
Threonine	5.72	5.80
Serine	6.42	6.56
Glutamic acid	12.66	13.45
Proline	4.62	4.62
Glycine	7.22	7.52
Alanine	8.20	8.22
Valine	6.33	6.00
Methionine	1.65	1.80
Isoleucine	4.65	4.57
Leucine	9.70	8.96
Tyrosine	2.78	2.69
Phenylalanine	4.25	4.28
Glucosamine	0.37	0.55
Galactosamine	1.35	1.71

[a] From Heller and Bok (1976a).

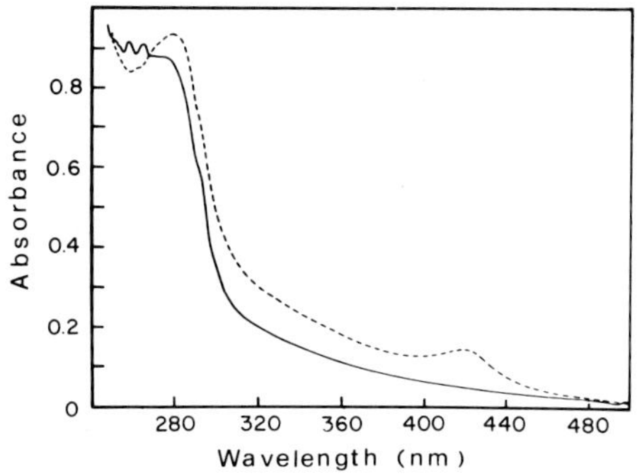

FIG. 13. Absorption spectra of purified pigment epithelial (-----) and rod outer segment fraction cytosol retinol-binding proteins. (From Heller and Bok, 1976a.)

VII. BINDING PROTEINS FOR RETINOL IN THE RETINA AND PHOTORECEPTOR CELL FRACTIONS

The possible existence of retinol-binding proteins in the retina has been investigated both on the whole retina tissue (Wiggert and Chader, 1975; Wiggert *et al.*, 1976; Futterman *et al.*, 1976; Saari and Futterman, 1976) and in isolated rod outer segments (Heller, 1976; Heller and Bok, 1976b). As for the binding proteins from pigment epithelium, the picture emerging from these first studies on retinal–cytosol-binding proteins is not uniform.

Utilizing incubation of whole retina cytosols with radioactive retinol and then centrifuging the material on a sucrose density gradient, a binding substance for retinol appearing as a single radioactivity peak with a sedimentation constant of 2 S and a molecular weight of 17,000 has been identified (Wiggert and Chader, 1975; Wiggert *et al.*, 1976) in the rat, guinea pig and chick embryo. Bovine retina cytosol has been reported by the same authors to contain, in addition to the 2 S binding species, a second more rapidly sedimenting substance that also showed the ability to bind exogenous radioactive retinol. The sedimentation velocity of this second radioactive peak was higher than that of the 5 S peak observed in serum samples and this could indicate that serum contamination was not responsible for it.

The 2 S binding substance was investigated in more detail and its presence confirmed by gel-filtration studies of radioactive retina cytosols. The binding of retinol has been reported to be specific (both retinoic acid and retinal do not appreciably compete with ^{3}H-retinol for the binding sites) and noncovalent. Its protein nature was inferred from the observation that the binding of retinol was substantially decreased by incubating the tissue cytosol with Pronase (100 μg/ml).

It should be noted that this 2 S receptor for retinol has been found by Wiggert and Chader in the retinas of 11-day-old chick embryos, appearing well before the development of the outer segments of the photoreceptor cells.

By fractionating on a DEAE-cellulose column the 100,000 *g* supernatant from bovine dark-adapted retina Saari and Futterman (1976) could identify five fractions showing an endogenous retinol-like fluorescence (Table IV). The fluorescence of fractions II and V appeared to be much more stable than that of the other fractions since it was only moderately decreased by extensive dialysis before the chromatography. Fractions IV and V were further purified by ion-exchange chromatography and investigated in some detail. Both fractions were shown by SDS gel electrophoresis to contain two polypeptides of very similar molecular weight. By both gel filtration and gel electrophoresis the molecular weight of purified fraction V was reported to be of the order of 17,000–17,500. No attempt to directly extract retinol from these fluorescent fractions was reported.

By incubating purified fraction V with human prealbumin, Saari and Futterman could demonstrate that about 40% of its total fluorescence binds to prealbumin; the authors excluded, however, that this could be the result of extensive contamination by serum RBP which, according to their calculations, could account for no more than about 5% of the total fluorescence of purified fraction V.

Only a part of fraction V, among the different fluorescent fractions of retina cytosol, was reported to bind radioactive exogenous retinol; recombination studies indicated that this radioactive part of fraction V does not bind human prealbumin.

The study of Saari and Futterman appears to be until now the only one

TABLE IV

Fluorescence Characteristics of Retinol-Binding Proteins from Bovine Retina Cytosol[a]

Fraction	Excitation λ max (nm)	Emission λ max (nm)
I	295–340	430
II	290–340	430
III	290–348	440
IV	290–345	443
V	295–350	470

[a] From Saari and Futterman (1976).

reporting the occurrence of an endogenous retinol-like fluorescence in the low-molecular-weight protein fraction of retina cytosol. In further investigations (Futterman *et al.*, 1976) these authors investigated in more detail some aspects of the binding specificity of this retinol binder and presented experimental evidence indicating that this soluble retinol-binding protein does not merely represent the apoprotein of serum RBP but is likely to be a separate protein molecule with possible physiological function. Binding of ^{3}H-retinol was reported to be specific and reversible. Binding studies with retinoic acid as well, by demonstrating the existence of different binding specificities for radioactive retinol and retinoic acid, were considered indicative of the existence in the low-molecular-weight fraction of retina cytosol of two separate binders.

In contrast with what is known to occcur with serum RBP, which does not easily exchange its chromophore, retinol-binding proteins for both retinol and retinoic acid have been reported (Futterman *et al.*, 1976) to readily release their bound radioactive molecule upon incubation with an excess of the specific unlabeled ligand.

A high-molecular-weight glycolipoprotein, very similar to that found by the same authors in pigment-epithelium cytosol, able to specifically bind exogenous radioactive retinol has been described by Heller (1976) and by Heller and Bok (1976b), in the cytosol from dark-adapted, bovine rod outer segments. As well as the pigment-epithelium, retinol-binding lipoprotein, the rod outer segment lipoprotein has been reported to have a molecular weight higher than 1.5×10^{6}. Its lipid content was found to be about 65% (by weight) total lipids and again the neutral lipid fraction contained only cholesterol and phosphatidylcholine accounted for over 90% of the phospholipid fraction. The protein is reported to contain at least four subunits on SDS gel electrophoresis (50,000, 75,000, 120,000 and 220,000 of molecular weight). Its absorption spectrum (Fig. 13) is different from that of the pigment-epithelium, cytosol–retinol-binding lipoprotein but the amino acid composition of the two proteins is practically identical (Table III).

VIII. CONSIDERATIONS AND SUMMARY

The literature reviewed in the preceding pages and the available experimental data reported therein allow some conclusions and justify some comments.

There seems to be little doubt that retinol is supplied to the retinal pigment epithelium only through the plasma RBP present in the extravascular space of the choriocapillaris and that for the active uptake of the vitamin the carrier plasma protein must interact with some receptor present on the basal plasma membranes of this monolayer of cells. Some evidence has been presented (Heller, 1975) that RBP does not penetrate the plasma membrane of pigment epithelium cells but is displaced from its membrane receptor after the release of the chromophore. The crucial event in this process seems to be the removal of the retinol molecule from the membrane-bound carrier plasma protein which as the apoprotein has a lower affinity for its membrane receptor than retinol-RBP.

Maraini *et al.* (1977) have reported that after the interaction of ^{3}H-retinol RBP with isolated pigment epithelium, the radioactive retinol is recovered bound to a 14,500-molecular-weight membrane protein probably different from circulating RBP and which they suggest might be a normal constituent of the plasma membranes of pigment epithelium. Whether this protein can represent the receptor for RBP and/or the receptor for retinol on pigment epithelium membranes is not presently known and further work is obviously needed to extend these data and to test these possibilities. If this protein is superficially placed in the membrane and exposed on its outer surface it would not be impossible by appropriate labeling techniques to investigate its distribution on the basal and/or apical cell surface and perhaps its motility in the membrane thus gaining further insight into its function.

As already said, the well-known instability of retinol in aqueous solution and its damaging effect on cell membranes (Bangham *et al.*, 1964) justified the search for cellular proteins that might help in solubilizing and protecting the chromophore, as happens during its transport in plasma (Futterman and Heller, 1972), and for proteins that perhaps act as carriers of the molecule in the exchanges known to occur between the pigment epithelium and the photoreceptors.

Similarly to what has been reported to occur in nonocular tissues, several such proteins have been described both in pigment epithelium and in the retina (Wiggert and Chader, 1975; Saari and Futterman, 1976; Heller, 1976). Many of them are low-molecular-weight proteins and at least the one described by Wiggert and Chader in the retina should not be exclusive of the photoreceptor's outer segments since it is present in the chick-embryo retina well before their appearance. However retinol-binding glycolipoproteins with molecular weights greater than 1.5×10^6 have also been isolated from both pigment epithelium and rod outer segments and partially characterized.

It should be stressed however that the only binding proteins showing endogenous retinol-like fluorescence are those isolated from bovine-retina cytosol according to Saari and Futterman (1976). All the other proteins so far described have been identified for their capacity to bind exogenous radioactive retinol when this was added to the tissue cytosol. Even if a specificity of binding was nearly always reported, this does only indicate that under particular experimental conditions a given protein is capable of binding retinol and does not necessarily demonstrate a physiological role for that molecule as a binding protein. The concentration of retinol in the tissue cytosol has been reported in some instances to be a crucial factor in those binding studies (Wiggert *et al.*, 1976) and the concentrations used in the experiment should not be necessarily considered as representative of the situation existing in the living cell.

An operational mechanism has been suggested (Heller and Bok, 1976b) that involves the possible role of these retinol-binding, high-molecular-weight lipoproteins as intercellular carriers of the chromophore from pigment epithelium to rod outer segments and back. This would imply the secretion into the interstitial space of the respective retinol-binding lipoprotein by the pigment epithelium cells and by the photoreceptor outer segments and the existence of specific receptors on the surface of the respective target cell. Though stimulating, this mechanism should be considered, as stressed by the authors themselves, only as a working hypothesis and other alternatives for retinol transport may be suggested. Even if some autoradiographic evidence suggesting a flow of phospholipid molecules from pigment epithelium to the photoreceptors has been presented by Bibb and Young (1974), no evidence at all has so far been presented that indicates that the retinol-binding lipoproteins may actually cross the outer membrane of the cell and act as intercellular carriers.

More convincing evidence of a possible physiological role for these proteins could possibly be obtained by experiments aimed at demonstrating that retinol is, in effect, the naturally occurring ligand of these binders under conditions stimulating a flow of retinol from the pigment epithelium to the photoreceptors or in the opposite direction. Investigations of pigment epithelial cells or rod-outer-segment extracts prepared from eyes in different conditions of light adaptation could perhaps be useful.

The clinical importance of these investigations does not need to be stressed. Both vitamin-A deficiency as well as retinol toxicity are known to cause the death of photoreceptor cells. A genetic deficiency causing a derangement of the molecular mechanisms of retinol utilization by pigment epithelium and/or the photoreceptors, or determining the intracellular accumulation of unbound retinol cannot a priori be excluded as a possible cause of the visual cell death that characterizes hereditary retinal degenerations.

Acknowledgments

I wish to thank Dr. M. Maione and Dr. G. L. Rossi for critically reading and discussing the manuscript. This work was supported in part by Grant 75.01056.04/115.2323. from Consiglio Nazionale delle Ricerche, Italy.

Addendum

Since submission of this manuscript to the Editor, many efforts have been directed to better characterize retinol-binding proteins in retina and pigment epithelium and a number of relevant papers have appeared in the literature. Some of them will be briefly mentioned here. Endogenous retinol-like fluorescence has been demonstrated in the 2 S receptor of chick-retina cytosol (Abe *et al.*, 1977), and compartmentalization of two distinct retinol receptors in the human retina is suggested by the lack of the 8 S species in the rodless retina of a patient with retinitis pigmentosa (Bergsma *et al.*, 1977). A cytosol protein binding specifically 11-*cis*-retinal has been described in human, rat, and bovine retina (Futterman and Saari, 1977; Futterman *et al.*, 1977). This receptor has a higher molecular weight than cellular retinol-binding protein and might be specific for the retina. Evidence has been obtained (Wiggert *et al.*, 1977; Saari *et al.*, 1977) indicating that cytosol retinol receptors in pigment epithelium and in the retina might be the same or closely similar proteins though the binder is enormously more concentrated (by a factor of more than 100) in pigment epithelium. Whole bovine retina extracts show approximately a 5-fold excess of retinoic acid-binding protein over retinol-binding protein. An interesting point, which could prove important in the correct interpretation of some experimental results, was the demonstration (Saari *et al.*, 1977) that preparations of isolated pigment epithelium consist almost entirely of cells with broken surface membranes and depeleted cytoplasm. The same conclusions have been reached in our laboratory during membrane-labeling investigations utilizing DANSCl ultrasonically dispersed into lecithin–cholesterol micelles (S. Ottonello and G. Maraini, in preparation).

References

Abe, T., Wiggert, B., Bergsma, D. R., and Chader, J. G. (1977). *Biochim. Biophys. Acta* **498,** 355–365.

Andrews, J. S., and Futterman, S. (1964). *J. Biol. Chem.* **239,** 4073–4076.

Bairati, A., and Orzalesi, N. (1963). *J. Ultrastruct. Res.* **9,** 484.

Bangham, A. D., Dingle, J. T., and Lucy, J. A. (1964). *Biochem. J.* **90,** 133–140.

Bashor, M. M., and Chytil, F. (1975). *Biochim. Biophys. Acta* **411,** 87–96.

Bashor, M. M., Toft, D. O., and Chytil, F. (1973). *Proc. Natl. Acad. Sci. U.S.A.* **70,** 3483–3487.

Baumann, Ch. (1972). *In* "Handbook of Sensory Physiology" (H. J. A., Datnall, ed.), Vol. VII/1, pp. 395–416. Springer-Verlag, Berlin, and New York.

Bergsma, D. R., Wiggert, B. N., Funahashi, M., Kuwabara, T., and Chader, G. J. (1977). *Nature (London)* **265,** 66–67.

Berman, E. R. (1964). *Biochim. Biophys. Acta* **83,** 371–373.

Berman, E. R., Schwell, H., and Feeney, L. (1974). *Invest. Ophthalmol.* **13,** 675–687.

Bibb, C., and Young, R. W. (1974). *J. Cell Biol.* **61,** 327–343.

Bill, A. (1976). *In* "Vision and Circulation" (J. S. Cant, ed.), pp. 333–338. Kimpton, London.
(J. S. Cant, ed.), pp. 333–338. Kimpton, London.
Bok, D., and Heller, J. (1976). *Exp. Eye Res.* **22,** 395–402.

Bridges, C. D. B. (1970). *In* "Biochemistry of the Eye" (C. N. Graymore, ed.), pp. 564–644. Academic Press, New York.

Bridges, C. D. B. (1975). *Vision Res.* **15,** 1311–1323.

Bridges, C. D. B. (1976a). *Exp. Eye Res.* **22,** 435–455.

Bridges, C. D. B. (1976b). *Nature (London)* **259,** 247–248.

Cone, R. A., and Brown, P. K. (1969). *Nature (London)* **221,** 818–820.

Cunha-Vaz, J. G., Shakib, M., and Ashton, N. (1966). *Brit. J. Ophthalmol.* **50,** 441–453.

Dowling, J. E. (1960). *Nature (London)* **188,** 114–118.

Dowling, J. E., and Wald, G. (1958). *Proc. Natl. Acad. Sci. U.S.A.* **44,** 648–661.

Fidge, N. H., Shiratori, T., Ganguly, J., and Goodman, DeW. S. (1968). *J. Lipid Res.* **9,** 103–109.

Futterman, S., and Andrews, J. S. (1964). *J. Biol. Chem.* **239,** 81–84.

Futterman, S., and Heller, J. (1972). *J. Biol. Chem.* **247,** 5168–5172.

Futterman, S., and Saari, J. C. (1977). *Invest. Ophthalmol. Visual Sci.* **16,** 768–771.

Futterman, S., Saari, J. C., and Swanso, D. E. (1976). *Exp. Eye Res.* **22,** 419–424.

Futterman, S., Saari, J. C., and Blair, S. (1977). *J. Biol. Chem.* **252,** 3267–3271.

Gambhir, K. K., and Ahluwalia, B. S. (1974). *Biochem. Biophys. Res. Comm.* **61,** 501–508.

Goodman, DeW. S., and Leslie, R. B. (1972). *Biochim. Biophys. Acta* **260,** 670–678.

Goodman, DeW. S., Huang, H. S., and Shiratori, T. (1965). *J. Lipid Res.* **6,** 390–396.

Heller, J. (1975a). *J. Biol. Chem.* **250,** 3613–3619.

Heller, J. (1975b). *J. Biol. Chem.* **250,** 6549–6554.

Heller, J. (1976). *J. Biol. Chem.* **251,** 2952–2957.

Heller, J., and Bok, D. (1976a). *Am. J. Ophthalmol.* **81,** 93–97.

Heller, J., and Bok, D. (1976b). *Exp. Eye Res.* **22,** 403–410.

Heller, J., and Horwitz, J. (1973). *J. Biol. Chem.* **248,** 6308–6316.

Heller, J., and Horwitz, J. (1974). *Exp. Eye Res.* **18,** 41–49.

Horwitz, J., and Heller, J. (1974). *J. Biol. Chem.* **249,** 7181–7185.

Huang, H. S., and Goodman, DeW. S. (1965). *J. Biol. Chem.* **240,** 2839–2844.

Hubbard, R. (1956). *J. Gen. Physiol.* **39,** 935–962.

Hubbard, R., and Colman, A. D. (1959). *Science* **130,** 977–978.

Hubbard, R., and Dowling, J. E. (1962). *Nature (London)* **193,** 341–343.

Hubbard, R., and Wald, G. (1952). *J. Gen. Physiol.* **36,** 269–315.

Kanai, M., Raz, A., and Goodman, DeW. S. (1968). *J. Clin. Invest.* **47,** 2025–2044.

Krinsky, N. I. (1958). *J. Biol. Chem.* **232,** 881–894.

Linder, M. C., Anderson, G. H., and Ascarelli, I. (1971). *J. Biol. Chem.* **246,** 5538–5540.

Lolley, R. N., and Schmidt, S. Y. (1974). *In* "The Eye" (H. Davson and L. T. Graham, Jr., eds.), Vol. VI, pp. 343–378. Academic Press, New York.

Maraini, G. (1974). *Invest. Ophthalmol.* **13,** 288–290.

Maraini, G., and Gozzoli, F. (1975). *Invest. Ophthalmol.* **14,** 785–787.

Maraini, G. and Gozzoli, F. (1976a). *In* "Vision and Circulation" (J. S. Cant ed.), pp. 107–112. Kimpton, London.

Maraini, G., and Gozzoli, F. (1976b). *Invest. Ophthalmol.* **15,** 1020–1022.

Maraini, G., Fadda, G., and Gozzoli, F. (1975). *Invest. Ophthalmol.* **14,** 236–237.

Maraini, G., Ottonello, S., Gozzoli, F., and Merli, A. (1977). *Nature (London)* **265,** 68–69.

Miller, S., and Steinberg, R. H. (1976). *Exp. Eye Res.* **23,** 177–190.

Morton, R. A. (1972). *In* "Handbook of Sensory Physiology" (H. J. A. Dartnall ed.), Vol. VII/1, pp. 35–68. Springer-Verlag, Berlin and New York.

Muto, Y., and Goodman, DeW. S. (1972). *J. Biol. Chem.* **247,** 2533–2544.

Muto, Y., Smith, F. R., and Goodman, DeW. S. (1973). *J. Lipid Res.* **14**, 525–532.
Noell, W. K., Delmelle, M. C., and Albrecht, R. (1971). *Science* **172**, 72–76.
Ong, D. E., and Chytil, F. (1974). *Biochem. Biophys. Res. Commun.* **59**, 221–229.
Ong, D. E., and Chytil, F. (1975). *J. Biol. Chem.* **250**, 6113–6117.
Peterson, P. A. (1971a). *J. Biol. Chem.* **246**, 34–43.
Peterson, P. A. (1971b). *Eur. J. Clin. Invest.* **1**, 437–444.
Peterson, P. A., and Rask, L. (1971). *J. Biol. Chem.* **246**, 7544–7550.
Peterson, P. A., Rask, L., Östberg, L., Andersson, L., Kamwendo, F., and Pertoft, H. (1973). *J. Biol. Chem.* **248**, 4009–4022.
Rask, L., Peterson, P. A., and Nilsson, S. F. (1971a). *J. Biol. Chem.* **246**, 6087–6097.
Rask, L., Vahlquist, A., and Peterson, P. A. (1971b). *J. Biol. Chem.* **246**, 6638–6646.
Rask, L., Peterson, P. A., and Björk, I. (1972). *Biochemistry* **11**, 264–268.
Raz, A., Shiratori, T., and Goodman, DeW. S. (1970). *J. Biol. Chem.* **245**, 1903–1912.
Saari, J. C., and Futterman, S. (1976). *Exp. Eye Res.* **22**, 425–433.
Saari, J. C., Bunt, A. M., Futterman, S., and Berman, E. R. (1977). *Invest. Ophthalmol. Visual Sci.* **16**, 797–806.
Sherman, B. S. (1970). *Exp. Eye Res.* **10**, 53–57.
Shichi, H. (1971). *J. Biol. Chem.* **246**, 6178–6182.
Shiose, Y. (1971). *Jpn. J. Ophthalmol.* **14**, 73–87.
Sidman, R. L., and Dowling, J. E. (1963). *Anat. Rec.* **145**, 286.
Smith, J. E., Muto, Y., Milch, P. O., and Goodman, DeW. S. (1973). *J. Biol. Chem.* **248**, 1544–1549.
Vahlquist, A., Nilsson, S. F., and Peterson, P. A. (1971). *Eur. J. Biochem.* **20**, 160–168.
Wald, G. (1933). *Nature (London)* **132**, 316–317.
Wald, G. (1934). *Nature (London)* **134**, 65.
Wald, G. (1935). *J. Gen Physiol.* **19**, 351–371.
Wald, G. (1937). *Nature (London)* **139**, 1017–1018.
Wald, G., Brown, P. K., and Smith, P. S. (1955). *J. Gen. Physiol.* **38**, 623–681.
White, G. H., Weston, S. M., and Glover, J. (1972) *FEBS Lett.* **27**, 107–110.
Wiggert, B. O., and Chader, G. J. (1975). *Exp. Eye Res.* **21**, 143–151.
Wiggert, B. O., Bergsma, D. R., and Chader, G. J. (1976). *Exp. Eye Res.* **22**, 411–418.
Young, R. W. (1967). *J. Cell Biol.* **33**, 61–72.
Young, R. W., and Bok, D. (1969). *J. Cell Biol.* **42**, 392–403.
Young, R. W., and Bok, D. (1970). *Invest. Ophthalmol.* **9**, 524–536.
Zadunaisky, J. A., and Degnan, K. J. (1976). *Exp. Eye Res.* **23**, 191–196.
Zimmerman, W. F., Yost, M. T., and Daemen, F. J. M. (1974). *Nature (London)* **250**, 66–67.

The Ocular Effects of Cannabinoids

KEITH GREEN[1]

I. Cannabinoid History . . . 175
II. Cannabinoid Chemistry . . . 178
A. Structure . . . 178
B. Metabolism . . . 179
III. Clinical Ophthalmic Tests . . . 181
A. Δ^9-Tetrahydrocannabinol in Normal Eyes . . . 181
B. Δ^9-Tetrahydrocannabinol in Glaucomatous Eyes . . . 184
C. Other Cannabinoids . . . 184
IV. Preclinical Tests . . . 185
A. Initial Studies: *in Vitro* . . . 185
B. Initial Studies: *in Vivo* . . . 186
C. Drug Delivery and Ocular Penetration . . . 190
V. Cannabinoid Pharmacology . . . 193
A. Theoretical Considerations . . . 194
B. Effect on Anterior Uveal Vessels . . . 195
C. Adrenergic Interaction with Tetrahydrocannabinol . . . 196
VI. Local and Central Action . . . 202
VII. New Cannabinoids . . . 204
A. Chemistry . . . 204
B. Pharmacology . . . 204
C. Ocular Effects . . . 205
D. Chronic Treatment . . . 208
VIII. Future Investigations . . . 210
References . . . 212

I. CANNABINOID HISTORY

The history and medicinal uses of cannabis, or marihuana as it is otherwise known, have been reviewed previously in detail by others

[1] Departments of Ophthalmology and Physiology, Medicial College of Georgia, Augusta, Georgia.

ISBN 0-12-153001-9

(Snyder, 1971; Nahas, 1973; Rubin, 1976) and will be only briefly reviewed here. According to Schultes (1973) *Cannabis* is one of the oldest plants known to man, and recorded history indicates that the first use of cannabis was over 2000 years ago, as a sedative (including induction of anesthesia) and all-purpose medication, in China. In early Indian literature the properties of the plant as a mind-altering substance were well known and its use was associated primarily with religious rites. The use of cannabis then spread to the Middle East, where the Arabic name for the *Cannabis* extract is hashish which means "grass." The known pleasure-inducing capacity of the plant was described at this time, but the use of cannabis as a pleasure-inducing, mind-altering drug did not spread to Western Europe until the latter half of the twentieth century. The plant extract, however, is still used widely as an all-purpose medication in India, South-East Asia, and many Caribbean islands for its known therapeutic effects.

Extracts of *Cannabis* were used throughout Europe as patent medicines to cure many diverse ailments, including headaches, rheumatism, and menstrual disorders. The highly variable potency of the extracts, however, gave extremes in therapeutic efficacy and, as more specific and uniform medications became available in the twentieth century, the *Cannabis* preparations fell into disuse. The medical use of cannabis in the United States, however, was eliminated by passage of the Marihuana Tax Act in 1937. From that time until the mid-1960s marihuana research was hampered by the lack of both specific compounds and scientific interest.

In the 1960s, however, the social use of marihuana entered the United States and, in that decade of social turmoil, reached epidemic proportions, becoming what is probably the most widely employed socially used drug, except for alcohol, in the history of Western civilization. The spread of marihuana as a drug of abuse took place over rather a short time span, at first, predominantly among the university student population, and second, down toward lower high school and junior high school grades. No factual knowledge existed concerning the drug, particularly with regard to its pharmacology or toxicology. Without any information upon which to base statements concerning the potential medical hazards of marihuana and fear that widespread marihuana use would lead to use of other drugs that were known to be harmful, the United States Federal Government began a research program on marihuana in 1967.

Other countries, notably Canada and Britain, have published long and comprehensive reports on both the scientific knowledge and social use of cannabis, which were the outcome of a long series of investigations. For the past 5 years the United States Department of Health, Education, and Welfare has also produced an annual Marihuana and Health report. The first report began with little factual information and much hearsay but the latest report contained objective presentations of data which were obtained under controlled conditions. The latest volume covered the effects of marihuana and its derivatives on a variety of activities in both clinical and preclinical tests.

Since the initiation of research programs that were conducted under scientific restraints, sponsored by the National Institute on Drug Abuse (NIDA) in the United States, much evidence has been produced concerning the biological changes induced by marihuana use. Reports have been made on the early phases of clinical studies of the acute pharmacological activities of tetrahydrocannabinol and of other cannabinoids (Hollister, 1971a,b; 1974). These reports revealed the primary characteristic acute responses of euphoria, tachycardia, and conjunctival redness. Studies of chronic marihuana users have, however, revealed effects on cellular metabolism, alteration of the immune response, and depression of male testosterone plasma levels, among others, which have all become areas of concern in marihuana research. It appears, however, that for each report of a pathological change caused by marihuana there is at least one other report that supports the opposite conclusion, namely that marihuana produces no change. Differences between the study groups, such as the length of time of marihuana use among the patients prior to the testing period, appeared to have a profound influence on the results that were obtained. Although there is reasonable evidence from laboratory studies to indicate that some changes occur in basic cell metabolism, hormonal regulation, and the immune system, the clinical significance of these observations is still not clear.

The metabolism of the main active ingredient of marihuana, tetrahydrocannabinol (THC) has also been the subject of intensive investigation over the past few years and some metabolites have been recovered over a week following a single exposure to THC. These new findings on the potential health hazard of marihuana use have led to the support of public programs aimed at discouraging the social use of this drug. Some of these reports have fueled the controversy about marihuana, with the publication of spectacular pathological findings. Many of the important findings are those which have been made during the last 2 years and thus require substantiation by others. The level of understanding of marihuana and its many effects is still low, but in some areas the degree of comprehension is growing rapidly.

Recent medical interest in the cannabinoids from a potentially therapeutic aspect has centered around the fact that these drugs have pharmacological activity profiles which render them unique among drugs, that they exhibit a very low toxicity in animals, as well as revealing no apparent physical dependence. Among the clear effects of marihuana are subjective sensations of pleasure, euphoria, and mild analgesia. Characteristically, conjunctival injection is normally present after marihuana use which indicates at least some effect of the drugs on the eye (Hollister, 1971b).

Since the determination of the structure of cannabidiol in 1963, advances in cannabis chemistry have been rapid, with the further synthesis of many structurally related compounds. The chemistry and pharmacology of the cannabinoids has become much better understood and new research utilizing pure, natural, or synthetic compounds has been undertaken over the past few years.

II. CANNABINOID CHEMISTRY

A. Structure

Prior to 1964 the major active component of cannabis had not been isolated nor its structure determined. Mechoulam (Gaoni and Mechoulam, 1964; Mechoulam and Gaoni, 1965) isolated and identified the active ingredient of marihuana, and also showed that the activity of Δ^9-tetrahydrocannabinol was not synergized by several major components of hashish (Mechoulam *et al.*, 1970).

The tetrahydrocannabinols are identified by two numbering systems; the first uses the rules for numbering pyran-type compounds and the second (Mechoulam, 1970) regards the compounds as substituted monoterpenoids. The structure (Fig. 1) of the primary active ingredient of marihuana is thus either Δ^9-THC or Δ^1-THC, respectively, depending on the classification that is followed. In this review the pyran system will be utilized unless otherwise stated. Δ^9-THC is lipid soluble and can be readily dissolved in oils; although other solubilizing agents such as the Tweens, polyethylene glycol or propylene glycol, and Triton have also been used, the result is usually an emulsion rather than a solubilized solution. In addition, propylene glycol has been shown to influence the metabolism and action of THC (Sprague and Craigmill, 1976). The vehicles themselves, therefore, may exert effects on the mode, or manner, of action of THC.

Sensitive methods have been developed to isolate and quantitate the various cannabinoids, and many compounds have been identified. Δ^9-THC has been shown to duplicate the effects of crude marihuana extract in animals and man, as does the isomer Δ^8-THC; the latter, however, is present in only small quantities ($<1\%$) in the plant extracts. Accompanying the enhanced detection methods for cannabinoids were advances in the chemistry of the synthesis procedures. The original synthesis of Δ^9-THC was long and expensive but methods have improved and currently synthetic Δ^9-THC is 95% pure but unstable when exposed to light and/or tempera-

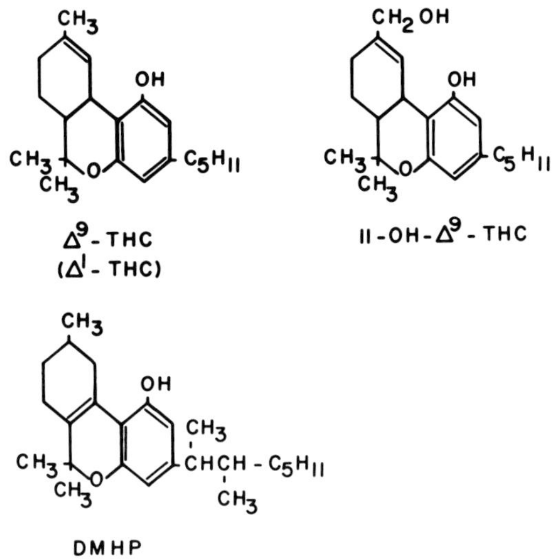

FIG. 1. Chemical structure of Δ⁹-tetrahydrocannabinol and some derivatives.

tures above 15°C. Many other cannabinoids have been synthesized under contracts from the NIDA and made available for research investigations. Several of these compounds have been tested for biological activity in animals and man (Hollister, 1974). The most potent analog was dimethylheptylpyran (DMHP) (Fig. 1) which was between 10 and 50 times more active than Δ^9-THC on an equimolar basis (Adams *et al.*, 1940). The cannabinoids are unique among agents that affect nervous activity in that they lack a nitrogen atom in the molecule, but other water-soluble and nitrogen-containing analogs have been synthesized and are pharmacologically active in many animal tests (Pars and Razdan, 1971; 1976).

B. Metabolism

Although Δ^9- and Δ^8-THC (Frankenheim *et al.*, 1971; Hollister and Gillespie, 1973) are known to cause the same effect as extracts of marihuana, it is not known whether the effects are caused by the parent compounds or some metabolite(s). In man and other animals Δ^9-THC is known to undergo a complex transformation which produces psychoactive and nonpsychoactive metabolites. Several of the metabolites appeared to be more active than THC in certain tests whereas some of these same metabolites were inactive relative to THC in other tests (Ben-Zvi *et al.*, 1970; Gill *et al.*, 1973; Mechoulam and Edery, 1973).

Δ^9-THC was rapidly metabolized *in vitro* by liver microsomes (Wall *et al.*, 1972; Ben-Zvi and Burstein, 1975) to 11-hydroxy-Δ^9-THC (7-hydroxy-Δ^1-THC) (Fig. 1), as well as cannabinol. After administration of radioactive Δ^9-THC or 11-hydroxy-Δ^9-THC, by intravenous injection or by smoking, about 50% of the radioactivity was found in the plasma, urine, and feces, with the latter being the major excretion route (Perez-Reyes *et al.*, 1973). Even after 72 hr about 50% of the original dose remained in body tissues. This remaining drug probably was slowly released from fatty tissues, with subsequent metabolism and excretion (Wall *et al.*, 1976). Initially, however, the Δ^9-THC was rapidly metabolized by hydroxylation to 11-hydroxy-Δ^9-THC, although other cannabinoids were also produced in small quantities. The plasma level of 11-hydroxy-Δ^9-THC was relatively small compared to the levels of Δ^9-THC (Wall *et al.*, 1976) and quantitatively amounted to about one-twentieth of that of Δ^9-THC at its peak. There was a further rapid conversion to an 11-carboxy acid form, which was one of the predominant metabolites; thus the conversion from Δ^9-THC to the 11-hydroxy form was a key metabolic step followed by conversion to the acid forms. This pathway appeared to be followed irrespective of the route of administration of Δ^9-THC. The 11-hydroxy forms were retained almost exclusively in the liver and excreted in the feces, which implied that they passed from the liver to the intestine via the gallbladder. The acid forms of the 11-hydroxy metabolites also occurred in large quantities and predominated in the plasma and urine.

Consideration has been given as to whether or not 11-hydroxy-Δ^9-THC is the active form of Δ^9-THC. This compound is more potent than Δ^9-THC on an equimolar basis but other metabolites have also been shown to be active. If the 11-hydroxy form was the only active compound it should be extremely potent, especially since the plasma concentration is low, but the degree of activity, relative to Δ^9-THC, is not greatly enhanced. Other sources for the appearance of the euphoric "high" and the physiological effects caused by THC were sought. Traces of cannabinol were also found to appear in plasma within 40 sec of Δ^9-THC administration (Mechoulam *et al.*, 1976b), which suggested that this cannabinoid might contribute to the psychoactive properties of Δ^9-THC. Cannabinol has, indeed, been shown to be active on several measures when given by intravenous injection (Wall *et al.*, 1976; Perez-Reyes *et al.*, 1976). Several metabolites have, therefore, been shown to exhibit psychoactive effects.

Despite the appearance of various metabolites after Δ^9-THC injection and the demonstration that they have psychoactive and other properties, time course studies suggest that cardiac rate changes (tachycardia) more closely parallel the levels of Δ^9-THC than the plasma metabolite concentration. There is a shorter latency of onset for some of the metabolites when they are injected directly, but the time difference is quite small. It is apparent, therefore, that the parent compound, as well as metabolites, contributes to the pharmacological activities of marihuana.

III. CLINICAL OPHTHALMIC TESTS

A. Δ^9-Tetrahydrocannabinol in Normal Eyes

It has been known for many years that one of the most obvious features resulting from the use of marihuana is conjunctival hyperemia. Other ophthalmic reactions were not measured or went unnoticed, except for the notion that users of marihuana were also reported to show a characteristically dilated pupil—also the end result of the use of other drugs.

Hepler and Frank (1971) reported the first objective measurements of the ophthalmic result of marihuana smoking in 1971 when they found that volunteers, who smoked 2 gm of marihuana (containing about 18 mg of THC) through an ice-cooled water pipe, showed a fall in intraocular pressure (IOP). The fall in IOP 1 hr after smoking ranged widely from one individual to another with a range from +4 to −45% from baseline levels with an average fall of 25%.

Hepler *et al.* (1972) made a complete ophthalmic examination of 21 adults prior to testing and the volunteers then smoked 2 gm of marihuana in a 30-min period. A noticeable hyperemia of the conjunctival blood vessels was seen reaching a peak about 30–60 min after smoking. Snellen visual acuity, refractive error, color vision, the degree of phoria, and stereoscopic perception were all unaltered. The pupil diameters were also unchanged, and the authors concluded that marihuana smoking did not decrease the pupillary reaction to strong light. Marihuana users do not, therefore, show a dilated pupil which is a characteristic response to many other drugs of abuse. The pupil diameters actually showed a tendency to decrease rather than increase under normal room light after smoking marihuana. In 10 eyes that were subjected to the Schirmer test, a constant and marked reduction in tear flow was found. The IOP was measured by applanation tonometry and the average reduction in IOP was 24% in this group of volunteers.

These findings have been substantiated by several authors after administration of Δ^9-THC either by smoking or intravenous injection. Adams *et al.* (1972) reported a 20% fall in the IOP of volunteers who smoked 12 mg of THC, and in a further study on 26 volunteers, also smoking 12 mg of THC, a similar decrease in IOP was observed (A. J. Adams, personal communication). In the latter study placebo was used as well as a crossover between placebo and drug. When marihuana was smoked there was a consistently lower IOP and a greater post-smoke decrease in IOP for marihuana compared to placebo. The maximum fall in IOP with marihuana occurred at 80 min with a 14.3% decrease, while with placebo a 4.6% fall in IOP occurred at 30 min. The IOP was measured with a low-displacement Goldmann applanation tonometer rather than the Schiøtz indentation tonometer, which allowed repeated measurements of IOP to be made without undue influence of the pressure measurements per se on the IOP. The pulse rate also increased, peaked at 5 min, and returned to normal by 30 min (prior to the time of maximal fall in IOP) but the "high" ranking, based on the subjective responses of the volunteers, was still at 70% of its maximum at this time.

More recent studies by this group (Flom *et al.*, 1975) indicated that the normal IOP in 15 volunteers fell only in those subjects whose experience was limited to moderate marihuana use, and for whom the decrease was associated with a "high" with a concurrent peaceful relaxation. The subjects smoked 12 mg THC, or a placebo (which consisted of extracted marihuana material) in a cigarette form. No blood pressure changes occurred with such low doses of THC, thus the fall in IOP could not be accounted for by a change in systemic blood pressure. There was a fall in IOP in 7 of 15 subjects who showed a significant correlation between the high and the relaxation tests. The authors contended that the reduced IOP was the result of a change in the psychophysiological status of the subjects who responded to THC, and suggested an indirect effect of the drug associated with relaxation and tiredness. In addition, it was found in these studies that individuals who used marihuana the most tended to have less fall in IOP; these subjects also showed little high or relaxation.

Purnell and Gregg (1975) also found that the maximal fall in IOP correlated with the peak effect of THC on the central nervous system, as indicated by the subjects report of their degree of anxiety. The hypotony, however, outlasted the anxiety feeling. Two healthy males were given THC intravenously after the method of Perez-Reyes *et al.* (1972), and the IOP was measured in both eyes of each subject with a Schiøtz indentation tonometer until the IOP returned to baseline. The IOP fell by about 52% in one subject, and by 43% in the other subject. The volunteers were titrated to peak euphoria using small increments of drug given by intravenous infusion and the maximum fall of IOP coincided with their euphoric maximum.

Cooler and Gregg (1976) presented further data on intravenous injection of Δ^9-THC in 10 volunteers and again found a significant fall in IOP which occurred in a dose-related manner. With a dose of 0.044 mg/kg body weight (approximately 3 mg total dose) the fall in IOP was 37%, and with 0.022 mg/kg body weight the fall in IOP was 29%, with a maximum reduction at both dose levels at about 90 min. Correlated with the fall in IOP was the feeling of anxiety as subjectively reported by the volunteers. This appears contradictory to the report of Flom *et al.* (1975) since in the studies of Gregg and co-workers the prevalent subjective response was anxiety rather than any relaxation effect of THC. The only common finding in the above studies was the fall in IOP and from the reports of others (Hepler *et al.*, 1976) it would not appear that relaxation is a common phenomenon in conjunction with the fall in IOP. It does appear from a consideration of all the available evidence that the fall in IOP is a THC-specific response.

Shapiro (1974) reviewed the ocular findings in 350 patients in Europe who used marihuana, either by smoking or orally. Both the objective and subjective manifestations of drug use were listed and the problems encountered were complaints regarding visual acuity, burning sensations, dryness of the globe, photophobia, and difficulty with accommodation, namely near vision. These problems appeared either as isolated signs or in combination. The IOP was quite low, averaging 10–12 mm Hg usually, with lows of 6–8 mm Hg occurring at times, which apparently held true of patients who denied drug use within the month preceding the measurement of IOP. Characteristically, however, it is quite common for patients to deny marihuana use despite recent smoking or ingestion of the drug. In view of the contrast between the data accumulated in controlled clinical environments, where other drug use was eliminated, and the data reported by Shapiro it would appear that interpretations of marihuana effects based upon the latter report should be viewed with caution.

One long-term study, that has been under way for the past 2 years in Costa Rica, contains an analysis of a battery of ophthalmic tests (Coggins, 1976). Of these tests only a decreased tear flow, as measured by the Schirmer test, and certain aspects of color vision have been found to be affected in a population of marihuana smokers versus a control group of cigarette smokers.

In another study, of 30 days' duration, in which twelve volunteers received high oral doses of THC (70 mg/24 hr in divided doses, which was increased to 210 mg/24 hr over a 1- to 5-day interval), some tolerance was noted in the IOP response to a smoked 1 gm marihuana cigarette (2.2% THC). At the beginning of the study there was a fall in IOP after cigarette smoking, but after ingestion of large oral doses there was no fall in IOP after cigarette smoking (Jones and Benowitz, 1976). It is unclear as to whether or not the reference to the lack of an IOP fall relates to the values of IOP which were found preceding the use of any drug or the values after oral ingestion but before smoking. The dose administered in one cigarette would only be a fraction (about 10%) of that given orally and one would predict that such a relatively minor increase in systemic drug concentration caused by smoking would be unlikely to produce any further response.

Controlled studies were reported by Hepler *et al.* (1976), who made examinations of 19 hospitalized subjects for 35 days and of 29 subjects for 94 days, all of whom had a normal IOP, who smoked a minimum of 2 marihuana cigarettes per day. The studies revealed the following: no changes in the pupil, visual acuity, depth perception or fusion, arterial blood flow, ophthalmodynamometry or electroretinography. The only ocular changes in this series (tear flow was not measured) were a consistent fall in IOP of 30% for about 4–5 hr following smoking of 2% marihuana cigarettes and a decrease in ocular pulse pressure during tonography.

B. Δ^9-Tetrahydrocannabinol in Glaucomatous Eyes

The first data on glaucoma patients were reported by Hepler *et al.* (1976). Eleven patients were studied, of which seven showed a fall in IOP averaging 30%. Further studies (Hepler and Petrus, 1976) on open-angle glaucoma patients, given THC either by smoking or oral ingestion, have shown that the fall in IOP averaged 29% with 1% THC (0.5 mg), 25% with 2% THC (1 mg), and 34% with 3% THC (1.5 mg), as measured 30 min after smoking. Following oral ingestion the IOP fall was 14, 23, and 24% with 5, 10, and 20 mg THC, respectively. Thus, irrespective of the route of administration, there was a dose–response effect of THC on the IOP, although greater quantities of THC were required orally than by smoking to elicit the same fall in IOP. The duration of the fall in IOP was about 4–5 hr whether the drug was given orally or by smoking. In addition, it was apparent that in those patients who could not be taken off their normal daily glaucoma medication during the tests, the effects of marihuana were additive to these drugs since an additional fall in IOP was caused by THC.

C. Other Cannabinoids

Perez-Reyes *et al.* (1976) examined the effects of various intravenously infused cannabinoids in volunteers. Three cannabinoids, namely, Δ^9-THC, 11-hydroxy-Δ^9-THC, and Δ^8-THC, produced a significant reduction in IOP. Correlated with the fall in IOP was the level of the high experienced by the subjects; although Δ^8-THC produced the greatest fall in IOP, the level of high experienced was significantly less than that with Δ^9- or 11-hydroxy-Δ^9-THC according to the subjective ranking of the volunteers. The effect on IOP also coincided with the psychological and cardiovascular effects that suggested that the fall in IOP may be a centrally mediated effect. The correlation, in almost all studies in man, between the subjective high (whether it be a relaxation or anxiety effect) and the fall in IOP must be more than coincidence, and the evidence is strong that the two effects coincide in time after drug administration. Such concurrent responses indicate that the fall in IOP may well be linked with a central nervous system response to the drug (see Section VI). That somewhat of a separation of effects exists is seen in the data where Δ^8-THC caused a marked reduction in IOP but had a lesser effect on the subjective euphoric experience, even though the total quantity of drug infused was greater than the other cannabinoids.

IV. PRECLINICAL TESTS

A. Initial Studies: *in Vitro*

Following the initial observation of Hepler and Frank in 1971 that there was a fall in IOP after marihuana smoking in man, experiments were begun in animals in an attempt to elucidate the mechanism of action of THC on the eye.

Initially, experiments were performed using an isolated, ciliary-body preparation and on anesthetized albino rabbits (Green and Pederson, 1973a). Fluid secretion across the isolated ciliary body of the rabbit eye was measured with no hydrostatic pressure between the two chambers on either side of the tissue, and fluid flows were measured with a transmembrane hydrostatic pressure from the stromal to the aqueous humor side, to determine passive hydraulic conductivity. When Δ^9-THC was applied to the ciliary body both secretion and short-circuit current, or active ion transport, were reduced by about 70 and 45%, respectively, at a concentration of 5×10^{-5} mg/ml. Concurrently, the electrical resistance of the membrane fell by 38% and the potential difference fell by about 60% (Table I). The concentration of THC used in the *in vitro* experiments was equivalent to that found in plasma of man 1 hr after smoking one marihuana cigarette (Galanter *et al.*, 1972). The same drug concentration increased the hydraulic conductivity by 70% over control values. That secretion was reduced by THC, as was the short-circuit current, indicated that THC affects the sodium transport system located in the membranes. The effect on fluid permeability and electrical resistance also suggested that there was a direct effect on the membrane.

Δ^9-THC has been shown to inhibit serotonin uptake, which is a transport system (Shaskan and Snyder, 1970), in rat brain synaptosomes (Sofia *et al.*, 1971). In addition, Porter and Scott (1970) observed an inhibition of cation transport in rat erythrocytes after THC treatment. Other studies have shown that THC increased the fluidization of liposomal cell membranes (Gill and Lawrence, 1976) as well as inhibiting synaptosomal ATPase activity (Poddar and Ghosh, 1976). Thus the findings on the ciliary

TABLE I

Permeability Changes Induced by Δ⁹-Tetrahydrocannabinol[a]

	In Vivo		*In Vitro*	
Dose (mg/ml)	Response	Value	Response	Value
Control	Protein	29.4 ± 6.5	Fluid	1.9 ± 0.2
50×10^{-6}	concentration (mg %)	56.6 ± 5.4	permeability (μl/mm Hg/min)	3.4 ± 0.4
Control	IOP	20.2 ± 0.9	Secretion	0.036 ± 0.003
50×10^{-6}	decrease (mm Hg)	15.1 ± 1.2	(μl/min)	0.010 ± 0.001
Control			Electrical resistance	60 ± 5
50×10^{-6}			(Ω cm^{-2})	38 ± 5

[a] Values are the mean ± SEM of at least 6 measurements.

epithelium are not surprising. Indeed, it is perhaps to be expected of a compound that shows such high lipid solubility that it would introduce itself into and between the lipid membrane components and cause a perturbation of cell membranes. Such membrane changes are of significance when considered in relation to the effects of THC on neuronal cells and the array of psychophysiological changes caused by THC. If THC alters cellular membrane conformation then it is entirely possible for permeability changes or changes in enzymic behavior to be an adjunct response to such structural alterations.

B. Initial Studies: *in Vivo*

1. Δ⁹-Tetrahydrocannabinol

Intravenous Δ⁹-THC (to give a concentration of 5×10^{-5} mg/ml of plasma) caused, in the urethane-anesthetized albino rabbit, a transient increase in heart rate of 15% at about 10 min, an increase of 100% in aqueous humor protein levels, an increase of 60–70% in total outflow facility (C_{tot}) and a fall in IOP, with a maximum of 25–30% after 60 min (Table I). Even in rabbits, however, there was a scatter of the data with a range of 10–60% as regards the fall in IOP (Green and Pederson, 1973a). Since only C_{tot} was measured the site of the increase of fluid flow from the eyes could not be determined, as C_{tot} is the summation of both true outflow facility through conventional pathways and pseudofacility (the suppression of aqueous humor inflow with increased IOP). Blood pressure remained unaffected by THC, as is the case in man. The THC was prepared using the solubilization procedure of Perez-Reyes *et al.* (1972) in which the THC was made into a microsuspension in 25% bovine serum albumin following initial solubilization in ethanol.

The fall in IOP of the rabbit was substantiated in part by the report of Cool *et al.* (1974). These authors measured the IOP of both albino and pigmented rabbits and administered Δ^9-THC both intravenously and topically. The IOP only fell in pigmented, immobilized rabbits, with approximately a 30% fall following a dose of 100–200 μg/kg body weight (approximately 30×10^{-4} mg/ml plasma) when the drug was given intravenously. The maximum decrease occurred at 50 min with a return to baseline IOP at 3 hr. The effect of topical drug application in either pigmented or albino rabbits was quite varied, however, although a fall in IOP was always found with either breed of rabbit. The range of fall in IOP was between 18 and 55%. Intravenous injections caused a concurrent reduction in systemic blood pressure at the same time as the fall in IOP, but topical administration caused no consistent changes in blood pressure.

Hepler *et al.* (1976) designed a delivery system capable of intoxicating albino rabbits with marihuana smoke. The fall in IOP was substantial (between 30 and 40%), as it was following the topical application to the eye of Δ^9-THC in 1% and 2% solutions in sesame oil.

Other studies (Dren, 1976) have also shown that Δ^9-THC lowered the IOP when applied topically to the albino rabbit eye as a suspension in 0.5% carboxymethylcellulose. A solution of 0.1% THC caused a fall in IOP of 30% at about 2 hr after application, with the fall in IOP lasting about 4 hr.

2. Prostaglandin Synthetase Inhibition

Further work confirmed the fall in IOP in albino rabbits following intravenous administration of Δ^9-THC (Green and Podos, 1974) to both the anesthetized and the awake, conscious animal. The IOP of the anesthetized animal was measured by direct manometry and in the conscious animal by use of the applanation tonometer (Walker and Litovitz, 1972). In addition, studies showed that Δ^9-THC inhibited the synthesis of prostaglandin from the precursor arachidonic acid when the latter was placed topically on the eye of a THC-treated animal. In a normal eye, topically applied arachidonic acid (2%) or prostaglandin E_2 (10 μg) caused immediate conjunctival hyperemia and engorgement of iris vessels, with a concurrent rise in IOP, an increase in C_{tot} and a 12-fold rise in aqueous humor protein levels. In the THC-treated (50×10^{-6} mg/ml plasma) animal after topical arachidonic acid, however, the eye remained quiet, the aqueous humor protein level increased only by about 100% and no substantial rise in IOP occurred (Green and Podos, 1974). THC had no influence on the ocular effects of topically administered prostaglandin E_2, thereby indicating that the THC effect was on the synthesis of prostaglandins from the precursor rather than on the action of prostaglandins themselves. THC has been shown in the ram seminal vesicle to inhibit prostaglandin formation from arachidonic acid (Burstein and Raz, 1972), but the dose levels of THC were high (0.1–10 μg/ml) compared to the experiments on the eye.

3. Other Cannabinoids

Experiments were performed to examine the relative efficacy of various intravenously administered THC metabolites or derivatives in reducing the IOP of the rabbit eye (Green and Bowman, 1976). The order of efficacy (see Table II) was 11-hydroxy-Δ^9-THC > Δ^9-THC > 8β-11-dihydroxy-Δ^9-THC > 8α-11-dihydroxy-Δ^9-THC > 8α-hydroxy-Δ^9-THC > all other derivatives that were tested. Compared to the order of potency given by Hollister (1974) for such derivatives in man, the two lists show a great deal in common, thus it would appear that the IOP responses in rabbit bear a great resemblance to other reactions seen in man. Cannabidiol had no effect on IOP and has little or no effect on other parameters of activity in man. The

TABLE II

Intravenous Marihuana Derivatives and Their Effect on Intraocular Pressure in the *in Vivo* Conscious Rabbit[a,b]

Drug[c]	Calculated plasma concentration (mg × 10^{-6}/ml plasma)	Maximal % fall in IOP
Δ^9-THC	50	26.4
	100	34.2
	200	37.2
11-OH-Δ^9-THC	50	28.6
	100	36.7
11-OH-Δ^8-THC	50–200	3.5
8α-OH-Δ^9-THC	50	2.2
	100–500	4.9
8α-11-diOH-Δ^9-THC	50	3.1
	100	11.1
8β-11-diOH-Δ^9-THC	50	3.7
	100	31.3
SP-111A	50	15.6
Cannabidiol	Inactive	

[a] Adapted from Green and Bowman (1976).

[b] Values are the mean of at least 6 measurements.

[c] Each drug was prepared according to the method of Perez-Reyes *et al.* (1972).

effect of topical administration of the derivatives in sesame oil was also tested and Δ^9-THC and 11-hydroxy-Δ^9-THC both produced a fall in IOP whereas other derivatives appeared to be inactive (Table III), at least at the concentrations tested. Of interest with the topical administration was the fact that there was a reduction of IOP in the contralateral eye, with a time course close to that in the eye receiving the drug.

Mechoulam and co-workers (Blumenthal *et al.*, 1976; Mechoulam *et al.*, 1976a) have described the effects of some cannabinoid derivatives in a rabbit glaucoma model. Eyes were treated with α-chymotrypsin injected into the posterior chamber and, after a few weeks, the IOP had increased to stable levels of 35 to 60 mm Hg. These animals then received medication into the conjunctival sac and the IOP was measured thereafter. The mode of delivery was quite unlike that of a drop, however, in that the lids were held open with the axis of the eye in the vertical position and a 0.5 ml solution held between the lid margins and overlying the cornea for 3 min. Obviously, there was no washout of drug by the tears, as occurs during the administration of a drop to the cornea, and the possibility for systemic absorption was very high. In addition, the cornea was exposed to the drug for a long period of time relative to the instillation of a drop, and the results might not be representative of a true *in vivo* situation.

Parenthetically, it might be added that the use of the rabbit for testing intraocular penetration of any drugs does not relate directly to the situation

TABLE III

Intraocular Pressure Fall in the *in Vivo* Conscious Rabbit after Topical Application of Δ^9-Tetrahydrocannabinol and Derivatives[a,b]

Drug	Dose (μg/50 μl)	Maximum % fall in IOP
Δ^9-THC	50	23.4
	100	28.9
11-OH-Δ^9-THC	50	26.4
SP-111A[c]	50	7.6
11-OH-Δ^8-THC	All inactive at 50 or 100 μg/50 μl	
8α-OH-Δ^8-THC	All inactive at 50 or 100 μg/50 μl	
8β-OH-Δ^9-THC	All inactive at 50 or 100 μg/50 μl	
8α-11-diOH-THC	All inactive at 50 or 100 μg/50 μl	
8β-11-diOH-THC	All inactive at 50 or 100 μg/50 μl	
Cannabidiol	All inactive at 50 or 100 μg/50 μl	

[a] Adapted from Green and Bowman (1976).

[b] Values are the mean of at least 6 measurements.

[c] The vehicle for all compounds was sesame seed oil, except for SP-111A where the vehicle was 0.9% saline containing 0.1% benzalkonium chloride. All volumes given were 50 μl applied directly to the corneal surface. Vehicle alone was applied to the contralateral eye.

in man because of the great difference in tear-film kinetics (Chrai *et al.*, 1973, 1974) especially as the blink rate of the rabbit (3–5 times per hour) is much slower than that of man (5 times per minute), although comparisons between vehicles for drug delivery do appear to offer valid extrapolation between these species. In addition, testing in the rabbit only indicates the IOP-lowering capability of a drug, not the psychoactive behavior, since the latter changes are difficult, at best, to assess in this species.

With their technique, Blumenthal *et al.* (1976) showed that Δ^6-THC (or Δ^8-THC in the pyran nomenclature) and cannabinol acetate produced a fall in IOP of long-standing duration. The fall in IOP was as long as those caused by norepinephrine and pilocarpine under the same experimental conditions, and dose levels as low as 0.01% were effective. It is of interest that Δ^8-THC was effective since it is the same compound that Perez-Reyes *et al.* (1976) concluded, from their studies in man, offers a good possibility as the drug of choice in the therapy of glaucoma, since Δ^8-THC has less psychoactive effects than Δ^9-THC in man, although causing a very significant fall in IOP.

C. Drug Delivery and Ocular Penetration

As is obvious from the preceding account, many techniques and vehicles have been used to administer Δ^9-THC and other derivatives directly to the eye. The problem is complicated because of the lipid solubility of the drugs and the difficulty in solubilizing them in solvents other than oils. Drug suspensions can be achieved that give reasonably satisfactory drug delivery, at least as measured by the biological response of a fall in IOP.

The delivery of Δ^9-THC (^{14}C-labeled) to the eye from various solvents has been measured (Green *et al.*, 1977b). The results are depicted in Fig. 2 where it can readily be seen that light mineral oil provided better delivery to the cornea and iris over a 6-hr time period than the other vehicles. The aqueous humor concentration, although lower, also reflected the same penetration pattern. Heavy mineral oil, sesame oil, and Tween 80/saline were less effective for drug delivery to all parts of the eye. The cornea appeared to act as a reservoir of THC for the remainder of the eye, since the corneal content was more than 10 times that of the iris and aqueous humor (on a total tissue weight basis).

Chromatographic analyses were not performed on the plasma samples after the use of labeled Δ^9-THC since the ^{14}C label is in a stable position and remains firmly attached to the molecule (Wall *et al.*, 1976). There must be some metabolism of Δ^9-THC since it has been reported that in rabbit, after intravenous injection, the half-life was 7–16 min (Agurell *et al.*, 1970; Teale *et al.*, 1975). The ^{14}C label, however, would remain on an active

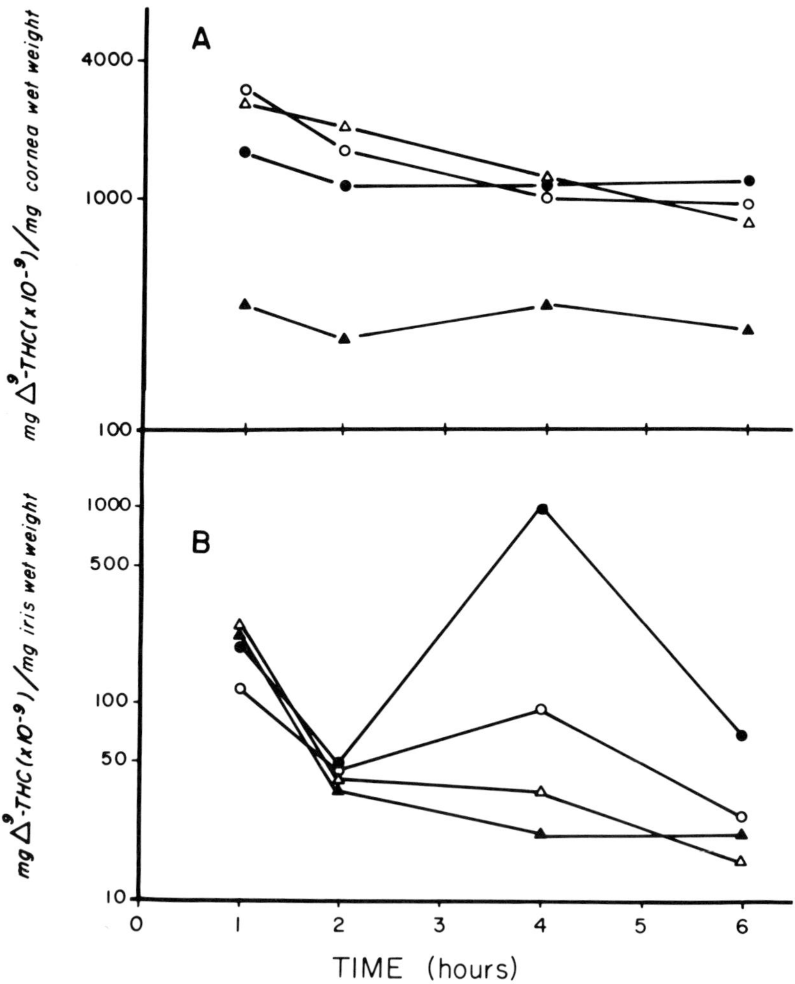

FIG. 2. Penetration of Δ^9-tetrahydrocannabinol into cornea and iris of the rabbit eye following topical application to the eye. Values are the mean of at least 6 measurements. (●) Light mineral oil, (○) heavy mineral oil, (▲) sesame oil, (△) 10% Tween–80/0.9% saline solution. (From Green *et al.*, 1977b.)

metabolite, presumably a more polar compound such as an 11-hydroxy form, rather than being lost from the molecule. The second peak concentration seen in the iris at 4 hr resembles that seen in the liver following intravenous injection of Δ^9-THC (Agurell *et al.*, 1970; Willinsky *et al.*, 1974). There are similarities between these tissues in terms of common types of organic transport systems (Bárány, 1973) and it may be that the second peak of activity represents an active tissue uptake of Δ^9-THC, by a transport system. Such a process would concentrate Δ^9-THC at a locus where it would, presumably, have its greatest effect (see Section V,B).

Since there are many varied mineral oils, we further examined four (Green *et al.*, 1977b) that had different viscosities, in terms of the biological response of reduced IOP to the dissolved THC. In this series, the least viscous oil (light, white mineral oil) proved to be most effective in delivering drug to the eye, as measured by the reduction of IOP.

The highest persistent dose of Δ^9-THC in the aqueous humor was 1×10^{-7} mg/ml, as determined from the penetration of labeled material, which is about 5 times less than the concentration producing a pronounced effect *in vitro* (Green and Pederson, 1973a). The drug was, therefore, reaching the target tissues via the locally administered route, i.e., the transcorneal pathway, and was present at concentrations sufficient to induce an effect on IOP (Green *et al.*, 1977a,b). Some experiments were performed with radiolabeled drug applied only to one eye and the data (Table IV) indicated that the aqueous humor and cornea of the contralateral eye contained very small quantities of drug. The iris, however, contained a concentration about 70% of that seen in the fellow eye that had received the topical drop. Substantial systemic absorption of the drug must have occurred, therefore, presumably through the conjunctival epithelium and vessels, coupled with a marked uptake mechanism in the iris from the systemic blood circulation. Both of these processes would, therefore, place the drug at loci where it could exert profound effects (see Section VI).

The mineral oil vehicle did not cause any irritation to the eye except a slight redness of the lids at the end of each day which cleared upon overnight cessation of drop application (Green *et al.*, 1977a,b). No epithelial defects were noted with light mineral oil unlike the other vehicles used. Sesame oil, heavy mineral oil, and the Tween 80/saline vehicle all caused epithelial defects of differing degrees which overtly stained with fluorescein.

TABLE IV

Concentration of ^{14}C-Labeled Δ^9-Tetrahydrocannabinol in Eyes of Rabbits Receiving Topical Application in One Eye[a,b]

	Aqueous humor	Cornea	Iris
Eye receiving drugs	41	1859	177
Contralateral eye	7	153	101

[a] Concentration of drug 1 hr after topical application.

[b] Values are the mean of at least 5 measurements. All values are mg ($\times 10^{-9}$)/mg tissue wet weight, except for aqueous humor values that are mg ($\times 10^{-9}$)/μl.

With sesame oil as the vehicle in earlier tests (Green and Bowman, 1976), the fall in IOP was quite variable, but with THC solubilized in mineral oil the fall in IOP was both more consistent and more extensive. It is of interest that the IOP fall was greatest with the lowest viscosity oil, since it has been shown in other studies that drug penetration and viscosity are not necessarily linked (Adler *et al.*, 1971; Green and Downs, 1975). Other aspects of the vehicle differences include the ability of the oil to release the drug, as well as its spreading capability within the conjunctival sac. The vehicle of choice, therefore, for topical application of lipid-soluble cannabinoids appears to be light mineral oil. Sesame oil has been widely used as a vehicle for THC, especially for oral use, but since this vehicle affects adrenergic synapses (Anton *et al.*, 1974) and causes corneal epithelium defects, it would appear prudent to replace sesame oil for topical ocular application of cannabinoids. The effect on adrenergic synapses assumes great significance when the pharmacological interactions between THC and the adrenergic nervous system are considered (Section V,B).

With regard to the water-soluble cannabinoids such as the new benzopyranopyridines (Pars and Razdan, 1976; Pars *et al.*, 1976), studies (Green *et al.*, 1977a,b) have indicated that a commercial tear film solution containing 1% methylcellulose and 0.01% benzalkonium chloride provided a good vehicle for these compounds when applied topically to the eye.

V. CANNABINOID PHARMACOLOGY

That there is a fall in IOP in man, monkey, and rabbit when THC is smoked, taken orally, given by intravenous infusion, or applied topically is a recognized fact, although all modes of delivery have not been tried in all species. What is less certain, however, is the mode of action of the drug as it affects the eye.

Both the experiments with radioactive Δ^9-THC (Section IV,C), as well as the physiological reduction in IOP of the contralateral eye (Green *et al.*, 1976, 1977a,b), indicated that significant systemic absorption occurred when drugs were applied topically. Sufficient drug was, therefore, present to cause a fall in IOP of the contralateral eye. Since systemic absorption is of such importance it is relevant to calculate the dose administered by this route: 50 μl of a 0.1% solution (50×10^{-3} mg) placed on the eye would result (assuming complete absorption) in a plasma concentration of 50×10^{-5} mg/ml (with a plasma volume of 100 ml) in a 3-kg rabbit or monkey with a plasma volume of 100 ml. This concentration is greater than that which was given intravenously in the initial rabbit studies (Green and

Pederson, 1973a). Complete absorption is, of course, unlikely but the plasma concentration is probably close to that achieved following systemic administration. Indeed, applying Δ^9-THC topically appears to result in a fall in IOP close to that found following intravenous injection (Fig. 3). Oral administration, however, requires a greater quantitative amount of cannabinoid in order to elicit a fall in IOP similar to that seen after a topical drop. Such a difference is probably related to the poorer absorption through the intestinal wall rather than destruction by the stomach or intestinal contents.

A. Theoretical Considerations

Both *in vitro* and *in vivo* THC increased the permeability of the ciliary epithelium (Table I). Assuming that all other parameters remain constant the IOP should naturally increase since more fluid would enter the eye. Since there is a fall in IOP, an explanation must be sought for these paradoxical findings of a fall in IOP in face of increased permeability of the membrane across which aqueous humor is formed.

The forces governing the inflow of aqueous humor are (Pederson and Green, 1973a)

$$L_p A \left\{x_c (P_c - P_o) - P_i\right\} + S$$

where L_p is the permeability of the ciliary epithelium (μl/min/mm Hg), A is the area (cm^2) of the membrane available for filtration, x_c is the dimensionless pressure index of the capillaries and the factor by which ophthalmic artery pressure must be multiplied to give the mean capillary pressure in

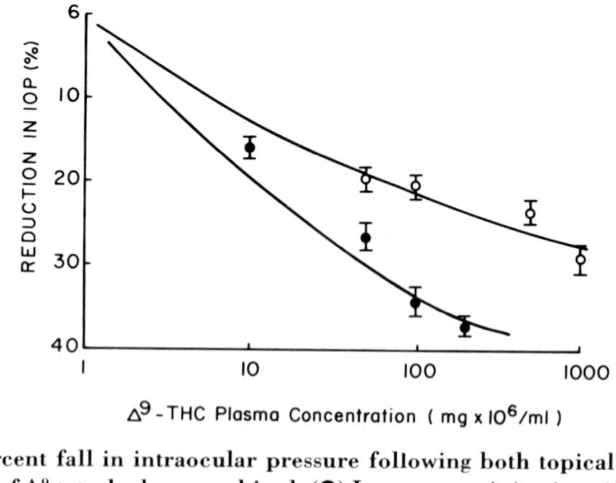

FIG. 3. Percent fall in intraocular pressure following both topical and intravenous administration of Δ^9-tetrahydrocannabinol. (●) Intravenous injection, (○) topical administration (50 μl).

the eye (Bárány, 1963), P_c is the capillary pressure in the ciliary body (mm Hg), P_0 is the plasma colloid osmotic pressure (mm Hg), P_i is the IOP (mm Hg), and S is secretion by the ciliary epithelium (μl/min).

Secretion, S, is the measure of fluid flow caused by active transport, which falls after THC treatment *in vitro*, and if secretion were solely responsible for aqueous humor formation then this reduction alone would account for the observed fall in IOP. Secretion is not the sole source of aqueous humor, however, as there is a significant fluid exchange between the ocular vasculature and the aqueous humor (Bárány, 1963; Pederson and Green, 1973b). Since secretion is a metabolically driven process it is independent of pressure gradients across the ciliary epithelium. It has been shown (Pederson and Green, 1973b) that secretion accounts for about 30–40% of aqueous humor formation with ultrafiltration accounting for the remainder. Aqueous humor inflow is, therefore, primarily dependent upon a net hydrostatic pressure driving force from the ciliary process capillaries to the aqueous humor (Green and Pederson, 1972, 1973b; Pederson and Green, 1973b).

The decrease in secretion does not, therefore, account for the fall in IOP following administration of THC. The fluid permeability, L_p, of the ciliary epithelium increased after THC and, as mentioned above, the IOP should naturally increase [the decrease in osmotic pressure difference between plasma and aqueous would change insignificantly; although the aqueous humor protein concentration increased by 100% from 30 to 50 mg%, the net difference remained almost unchanged since the plasma concentration was high (4500 mg%) relative to that of the aqueous humor].

To account for a fall in IOP, therefore, other changes must be occurring. Other factors in the equation subject to change are P_c, A, and x_c. P_i is IOP and obviously is decreased, and P_0 is constant throughout these experiments. It is unlikely that the area alone would change, without a concomitant change in P_c or x_c. The driving force for ultrafiltration is the pressure differential between the capillaries of the ciliary processes and the aqueous humor, and for less ultrafiltrative flow this pressure difference must be reduced. From the available data, therefore, one must conclude that P_c should fall either by vasoconstriction of the afferent vessels, which would affect both x_c and P_c, or by vasodilation of the efferent vessels which would reduce P_c and x_c. In any event, a reduction in capillary pressure would follow with a decrease in the rate of ultrafiltration across the ciliary epithelium in face of an increase in the permeability of the membrane.

B. Effect on Anterior Uveal Vessels

In support of such an effect of THC on anterior uveal capillary blood pressure are data obtained on eyes subjected to paracentesis, which normally causes an initial hypertensive phase in IOP. Administration of intravenous Δ^9-THC prevented this hypertensive phase. In this regard, Δ^9-THC has an effect similar to that of epinephrine-like compounds (Leopold and Day, 1950, 1951) and Diamox (Leopold, 1955; Leopold *et al.*, 1955; Kronfeld and Freeman, 1960) which also prevented the primary and secondary rises in IOP after paracentesis. Paracentesis is known to cause the formation of prostaglandins by the iris and anterior uveal tissue which in turn induces vasodilation of the afferent vessels of the anterior uvea (Whitelocke and Eakins, 1973).

Leopold and Day (1950) attributed the ocular response to the fact that the epinephrine-like compounds could induce vasoconstriction of the afferent uveal vessels which directly antagonized the vasodilatory stimulus of paracentesis by acting on the same vessels. In a similar context, it appears that THC, by means of its vasodilatory capacity on the efferent vessels, acts to physiologically negate the IOP-increasing effect of increased capillary pressure caused by paracentesis-induced vasodilation of the afferent vessels. The drug action resembles the effects seen by Macri and Cevario (1974a) with different drugs causing either increased or decreased aqueous humor formation by acting on different vascular loci. Certainly THC antagonizes the normal ocular response to prostaglandin E_2 (PGE_2) infusion, (0.079 μg/7.9 μl/min), sufficient to cause a 5 mm Hg elevation of IOP in a normal eye (Green and Kim, 1975) as evidenced by the failure of the IOP to increase in response to PGE_2 following intravenous Δ^9-THC administration 30 min prior to PGE_2 (K. Green and K. Kim, unpublished observations).

C. Adrenergic Interaction with Tetrahydrocannabinol

The acute pharmacological response to THC is a fall in IOP with a concurrent increase in C_{tot} (Green and Pederson, 1973a). The mechanism of this action was poorly understood, however, until studies were made following interference with adrenergic pathways to the eye and the interaction between these pathways and THC.

Current concepts of the influence of adrenergic innervation of aqueous-humor dynamics in the rabbit involve the presence of α-receptors which have unequivocally been demonstrated to be responsible for the regulation of C_{tot}, with a small β-adrenergic effect (Sears and Neufeld, 1975). Detailed studies have shown that true outflow facility is exclusively controlled by α-adrenergic innervation systems (Eakins, 1963), which implies that the minor β-adrenergic effect would be on pseudofacility. With regard to aqueous humor formation, both the α- and β-adrenergic receptors have significant roles. Stimulation of α-receptors produces a minor increase in aqueous humor formation and stimulation of β-receptors produces a major fall in aqueous humor formation (Chiou and Zimmerman, 1975). Both of these latter two effector systems may well exert their control on blood flow regulation through the ciliary body, thus influencing aqueous humor formation by this route rather than having a direct effect in the ciliary epithelium per se (Macri and Cevario, 1974a,b). There is a small effect of α- and β-adrenergic agonists on the *in vitro* ciliary epithelium (Pederson and Green, 1973b; K. Green and C. Griffin, unpublished observations), but the most pronounced effect is probably on the vasculature of the anterior uvea.

1. Adrenergic Antagonists

Both α- and β-adrenergic antagonists were utilized in an attempt to discern the site of action of Δ^9-THC in the eye (Green and Kim, 1976a). Both α- and β-adrenergic antagonists inhibited, to different degrees, the THC-induced increase in C_{tot} and the decrease in IOP (Table V). The α-adrenergic antagonists completely blocked the THC-induced increase in C_{tot}, and the β-adrenergic antagonists blocked the decrease in IOP.

It was suggested (Green and Pederson, 1973a) that the THC-induced increase in C_{tot} was due, in part, to an increase in pseudofacility since the ciliary epithelium permeability was increased. The α-adrenergic antagonist inhibition of the increase in C_{tot} indicates that THC acts by increasing true outflow facility and, conversely the lack of any β-adrenergic antagonist effect suggests that β-adrenergic receptors play no role in regulating this response to THC. It was thought that the α-adrenergic antagonists may inhibit a THC-induced effect on episcleral blood vessels involved with aqueous humor drainage and thereby regulate outflow facility in this manner. It is apparent from Table V that the fall in IOP caused by THC was reduced by a half following the use of the

TABLE V

Intraocular Pressure Change Caused by Adrenergic Antagonist Alone and in Combination with THC in the Normal Eye[a]

Drug	Percentage fall in IOP[b]	(Antagonist + THC) − (antagonist alone)
THC alone	22.5	
Phenoxybenzamine (0.1 mg/kg body wt.)	11.4	7.3
Phenoxybenzamine + THC	18.7	
Sotalol (0.5 mg/kg body wt.)	7.2	−3.5
Sotalol + THC	3.7	

[a] Adapted from Green and Kim (1976a).

[b] The percentage fall in IOP reflects the change in IOP between 30 and 90 min, since THC was administered 30 min after adrenergic antagonist.

α-adrenergic antagonist phenoxybenzamine, whereas the β-adrenergic antagonist, sotalol, inhibited the THC-induced fall in IOP completely.

These findings may be related to those seen with drugs in the isolated, arterially perfused cat eye preparation (Macri and Cevario, 1974a,b). These authors found that drugs caused different effects on aqueous humor formation depending upon the site of action in the vessels leading to (afferent) or from (efferent) the ciliary body. With vasoconstriction of the afferent vessels the rate of aqueous humor formation decreased, as indeed it did when the efferent vessels vasodilated. In contradistinction, vasodilation of the afferent vessels and vasoconstriction of the efferent vessels increased aqueous humor formation. The implication of this work is that both the afferent and efferent vessels are capable of exerting a great influence on blood flow through the anterior uvea. The venous side, therefore, should be capable of influencing blood flow to an almost equal extent as the arteriolar side (Macri and Cevario, 1974a).

Since THC reduced capillary pressure in the eye (Section V,A and B), the results obtained indicate that THC acts in the normal eye by vasodilating the efferent blood vessels since a β-adrenergic antagonist inhibited the THC effect completely (Table V), a process that would reduce the capillary pressure in the ciliary body and cause a fall in IOP (Green *et al.*, 1978). Such an action is, in addition, consistent with the known vasodilatory action on conjunctival vessels (Helper *et al.*, 1972), an effect that is also inhibited by β-adrenergic blockade (Vachon and Sulkowski, 1976). The effect of other antagonists on the THC-induced reduction in IOP may also relate to a possible differential action on the ciliary-body blood supply. From the experimental data with the interaction between THC and adrenergic antagonists it was obvious that some mediation of the IOP response occurred through adrenergic receptors.

Further elucidation of the site and mode of action of THC was performed in the normal rabbit eye (Green and Kim, 1976b), and the experiments confirmed that the THC effect was mediated by the adrenergic nervous system. Initially, a ganglionic blocking agent, hexamethonium chloride (1 mg/kg body weight), was employed to block ganglionic transmission. Hexamethonium chloride alone caused a small (9%) fall in IOP which was accompanied by a minor fall in systemic blood pressure and an increase of 15% in C_{tot} (Table VI). THC and hexamethonium chloride in combination caused a greater fall in IOP but the change was only 68% of that induced by THC alone. Assuming that the drug effects are purely additive, THC, in the presence of hexamethonium chloride, caused only a 32% fall in IOP (Table VI). In the presence of hexamethonium chloride, THC had a much reduced effect on IOP but an almost normal effect on C_{tot}. Ganglionic blockade appears, therefore, to only inhibit the THC-induced fall in IOP

TABLE VI

Effect of Δ^9-Tetrahydrocannabinol on Aqueous Humor Dynamics after Adrenergic Blockade or Ganglionectomy[a,b]

Drug[c]	n[d]	Fall in IOP[e]	Fall in IOP as % of THC effect	Rise in C_{tot}[f]	Rise in C_{tot} as % of THC effect
THC	6	25.23 ± 3.41	100	24.89 ± 2.91	100
THC + gangx	6	9.67 ± 1.75	38	14.13 ± 2.58	57
Hexamethonium chloride	6	8.98 ± 2.42	36	15.25 ± 3.24	60
Hexamethonium chloride + THC	6	17.39 ± 2.99	68	34.67 ± 3.87	140
THC alone (line 4 minus line 3)		8.41	32	19.42	80

[a] Adapted from Green and Kim (1976b).

[b] All values are the mean ± SEM.

[c] Drug, THC, was injected to give 50×10^{-6} mg/ml plasma; hexamethonium chloride was given at 1 mg/kg body weight; gangx, superior cervical ganglionectomy.

[d] n, Number of experiments.

[e] Expressed as $\frac{IOP_0 - IOP_{60}}{IOP_0} \times 100$.

[f] Expressed as $\frac{C_{tot_{60}} - C_{tot_0}}{C_{tot_{60}}} \times 100$.

without affecting the effect on C_{tot}, indicating that for the revelation of the IOP effect a connection to the central nervous system is necessary while for the C_{tot} effect, such a connection is not required.

2. Ganglionectomy

Use of hexamethonium chloride resulted in both eyes of an animal receiving the same treatment, thus it was difficult to examine the effects of THC in normal eyes and those in which the adrenergic supply was inhibited in the same animal. This problem was overcome by performing unilateral superior cervical ganglionectomy on rabbits. In this case, the adrenergic innervation to one eye was deleted since the adrenergic nerves within the eye resorb within a 14-day period (Roth and Richardson, 1969) following removal of the ganglion. Animals were only used at least 10 days postsurgery and after demonstration of supersensitivity to 0.1% norepinephrine.

Intravenous THC at a plasma concentration of 50×10^{-6} mg/ml plasma caused a 25% reduction in IOP together with a 25% increase in C_{tot} (Table VI) in the eyes of normal animals, or the normal eye of unilaterally ganglionectomized animals. The fall in IOP of the ganglionectomized eye was only 38% of that in the contralateral normal eye and the increase in C_{tot} was only 30% of the normal THC-induced increase (Table VI). The fall in IOP and rise in C_{tot} found in the ganglionectomized eye is a reflection of the peripheral sympathomimetic action of THC, since there can be no transfer of a central stimulus (Green *et al.*, 1976). To further comprehend the effects of THC in the eye it is revealing to compare and contrast the similarities and differences between the hexamethonium chloride-treated and the ganglionectomized eye. The effect on IOP, although reduced, must still be the result of some degree of dilation induced in the efferent anterior uveal vessels in both hexamethonium chloride-treated and ganglionectomized eyes. In the ganglionectomized eye, with no central connection via the cervical ganglion, both the IOP and C_{tot} effects of THC are substantially reduced, whereas in the presence of a ganglionic blocking agent only the effect of IOP is suppressed (Table VI). The reduced effect of THC on IOP in both hexamethonium chloride-treated animals and ganglionectomized eyes of normal rabbits strongly indicates that, apart from direct peripheral sympathomimetic actions, there is an effect exerted through the central nervous system either at ganglionic or preganglionic sites, to evoke the normal respone to THC. Ganglionic blockade and ganglionic sympathectomy both interrupt the normal passage of impulses through the adrenergic pathways, but only after ganglionectomy is there loss of norepinephrine stores in the eye that potentially can be released to increase C_{tot} in the normal eye.

The lessened effect on C_{tot}, compared to the effect in the normal or hexamethonium chloride-treated eye, could be due to the fact that the norepinephrine stores in the eye are depleted following nerve degeneration after ganglionectomy (Eakins and Eakins, 1964; Roth and Richardson, 1969). The small, reduced effect of THC on C_{tot} in the ganglionectomized animals could, therefore, be the result of a pure α-agonist effect of THC. In the presence of hexamethonium chloride, however, since norepinephrine is still present in the nerve terminals, THC can enhance its release by inhibiting prostaglandin synthesis. Together THC and norepinephrine act synergistically, both as α-agonists, on C_{tot}, as in the normal eye, to produce the "normal" THC-induced rise in C_{tot}. In favor of this interpretation is the α-adrenergic antagonist inhibition of the THC-induced effect on C_{tot} (Section V, C, 1).

Neufeld and Page (1975) have provided a scheme of adrenergic interactions in the eye that includes prostaglandin synthetase located in the postsynaptic junction. It is known from the work of Hedqvist (1973; Hedqvist and Brundin, 1969) that prostaglandins modulate the release of norepinephrine in the synapses of several tissues, and that inhibition of prostaglandin synthesis leads to more presynaptic norepinephrine release.

Since THC is a prostaglandin synthetase inhibitor (Burstein and Raz, 1972; Burstein, 1976; Green and Podos, 1974) the presence of this drug would result in more norepinephrine being released from the nerve synapses. The quantity of norepinephrine released, however, cannot be great since no mydriasis is noted following THC administration (Green *et al.*, 1977a), but perhaps sufficient norepinephrine is released to increase C_{tot}. It is known that the two effects, mydriasis and an increase in C_{tot}, can be dissociated in terms of the effect of varying norepinephrine concentrations (Eakins and Eakins, 1964) and the present data probably indicates a difference in the quantitative effects of two separate systems.

3. Parasympathetic Pathways

A parasympathetic effect of THC can be ruled out since both hexamethonium chloride, which blocks both sympathetic and parasympathetic transmission at ganglia, produces the same reduction of IOP as does ganglionectomy. Further evidence for the lack of involvement of the parasympathetic innervation system was obtained from studies using atropine (Green *et al.*, 1977a) and scopolamine (K. Green and K. Kim, unpublished observations) drops administered topically prior to the injection of THC. The fall in IOP with either drug in the presence of THC was the same as with THC alone. The rabbit destroys atropine rapidly (Havener, 1974) but scopolamine is not metabolized rapidly; neither drug, however, caused any modulation of the normal THC response, which indicates that the parasympathetic pathways are not involved in the mediation of the THC response.

4. Optic Nerve-Sectioned Rabbits

The mechanism of the response of THC in the ganglionectomized eye could potentially have been mediated through some possible receptor system such as that seen previously in optic nerve-sectioned rabbits with various drugs where a receptor system with a central nervous system connection via the optic nerve has been demonstrated (Podos *et al.*, 1970). No statistically significant difference was found between the reduction in IOP in the normal or optic nerve-transected eye following topical application of Δ^9-THC. These results occurred irrespective of the eye into which the drop was placed. Optic nerve transection, therefore, did not block the ocular response to THC (Green *et al.*, 1976, 1977b).

5. Adrenergic Antagonists in Ganglionectomized Eye

The local sympathomimetic action of THC on IOP in the ganglionectomized eye can be effectively blocked using the β-adrenergic antagonist, sotalol (Table VII). In addition, the THC-induced fall in IOP was partially antagonized by the α-adrenergic antagonist, phenoxybenzamine;

TABLE VII

Intraocular Pressure Change Caused by Adrenergic Antagonist Alone or in Combination with THC in the Ganglionectomized Eye

Drug	Percentage fall in IOP[a]	(Antagonist + THC) − (antagonist alone)
THC	10.2	
Phenoxybenzamine	1.5	
(0.1 mg/kg body wt.)		4.3
Phenoxybenzamine + THC	5.8	
Sotalol (0.5 mg/kg body wt.)	3.2	0.3
Sotalol + THC	3.5	

[a] The percentage fall in IOP reflects the change in IOP between 30 and 90 min, since THC was administered 30 min after adrenergic antagonist.

the latter drug also decreased the THC-induced increase in C_{tot}. This suggests that THC does possess α- as well as β-adrenergic activity in the eye. The central nature of the stimulus is illustrated by a comparison of preganglionectomized and ganglionectomized eyes. In normal eyes the fall in IOP elicited by THC is about 25% and in the ganglionectomized eye is about 9%. Preganglionectomy (removal of a 1 cm segment of the nerve about 1.5 cm away from the ganglion) caused a response that was roughly halfway between, with a fall in IOP of about 17% (Green *et al.*, 1976) which illustrates that the pathway from the central nervous system is vital for the full revelation of the THC effect. That there is a possible effect of THC on ganglionic sites is illustrated by the difference in the fall in IOP of ganglionectomized and preganglionectomized eyes.

VI. LOCAL AND CENTRAL ACTION

Topical THC application to unilaterally ganglionectomized rabbits also revealed that systemic absorption occurred, as illustrated by the following experiments. THC, dissolved in mineral oil, was applied topically (50 μl of a 0.1% solution) to either normal or denervated eyes and the IOP was measured with an applanation tonometer. Addition of THC to a normal eye caused a fall in IOP of 25% after 1 hr, but the IOP in the contralateral ganglionectomized eye only fell by 7% after the same time (Green and Kim, 1976b). Application of THC to a ganglionectomized eye caused the same changes, that is, the IOP of the contralateral normal eye fell by 25% and that of the ganglionectomized eye by 7%. This again illustrates the central nature of THC in reducing the IOP and reveals that systemic absorption plays a major role in the action of THC (Table VIII), supporting the

TABLE VIII

Indications for Local and Systemic Actions of THC

Local	Systemic
(a) Penetration of ^{14}C-THC into eye to levels sufficient to affect ciliary epithelium (b) Adrenergic receptor blocking agents inhibit THC-induced fall in IOP	(a) Uptake by contralateral iris of ^{14}C-THC administered to one eye (b) IOP falls in both eyes irrespective of the placement of a topical drop, with a similar time course (c) Ganglionectomy and preganglionectomy reduce the THC-induced fall in IOP

previous evidence obtained from the iris uptake of systemic radioactive THC and the fact that the IOP falls in both normal eyes of the same animal following drug application to only one eye. The purely peripheral sympathomimetic action of THC on IOP is, therefore, responsible for about 30% of the fall in IOP and most of the remainder is caused by central nervous system stimulation.

That marihuana, or THC, may affect central nervous system loci has been demonstrated by other authors (Cavero *et al.*, 1972; Vollmer *et al.*, 1974; Pars and Razdan, 1976; Vachon and Sulkowski, 1976). Pars and Razdan (1976) showed that β-adrenergic stimulation could account, in part, for the behavioral effect seen with several of the newer cannabinoids (Pars *et al.*, 1976; Razdan *et al.*, 1976). In the cat, use of the β-adrenergic antagonist dichloroisoproterenol prevented many of the depressant effects of the cannabinoids that were tested. One of the new synthetic cannabinoids (see Section VII) also potentiated the pressor response of both epinephrine and norepinephrine (Dewey *et al.*, 1970; Zitko *et al.*, 1972). Other studies have indicated that several cannabinoids inhibit the uptake of norepinephrine in various regions of the rat brain, particularly in hypothalamic synaptosomes (Banerjee *et al.*, 1975). In the cat, the cardiovascular effects of Δ^9-THC appear to be due to a central nervous system action which results in a decrease in sympathetic tone (Vollmer *et al.*, 1974).

When volunteers smoked marihuana, the usual responses were observed (Vachon and Sulkowski, 1976) but pretreatment with propranolol, a β-adrenergic antagonist, prevented the tachycardia and rise in blood pressure as well as markedly reducing the degree of conjunctival injection. In addition, reduced effectiveness of marihuana was noted with several performance measures and in the subjective high that was experienced following pretreatment with propranolol. Vachon and Sulkowski (1976) concluded that the effect of propranolol was due to either peripheral or central actions. β-Blockade prevention of slight memory impairment, and other performance changes normally induced by marihuana, should be assigned to the interaction of propranolol and marihuana in the central nervous system, however, and the authors concluded that adrenergic pathways were involved in the production of some marihuana effects.

Obviously, the fact that THC acts centrally to produce effects on electroencephalograms and other parameters of brain activity, particularly behavioral effects, does not prove that THC acts on the eye via an adrenergic pathway, but the evidence is convincing that some centrally located site profoundly influences the eye when stimulated by THC.

VII. NEW CANNABINOIDS

That Δ^9-THC is effective in reducing IOP is undisputed (Green, 1975) whether delivered in low doses either intravenously, topically, orally, or by smoking. For potential clinical use, however, a topical drop form of one of the cannabinoids or another related compound is more desirable since unacceptable psychotropic action could be avoided and since it is readily self-administered.

A. Chemistry

Recently, newer nitrogen-containing analogs of the cannabinoids, nitrogen-substituted benzopyranopyridines, which are historically and structurally related to the cannabinoids, have been synthesized (Fig. 4), and have similar activity to the parent compounds in many pharmacological tests (Pars *et al.*, 1976; Pars and Razdan, 1976; Razdan *et al.*, 1976). SP-1 is an oil-soluble, nitrogen-containing compound with the form of a white crystalline solid that is oil soluble but easier to handle than the resinous Δ^9-THC. Another compound, SP-106 (Fig. 4), has an ester group on the phenolic hydroxyl that confers water solubility on the compound. A further compound, SP-204, has a methyl group on the phenolic group and is also a basic water-soluble ester. These drugs were synthesized and supplied by SISA, Inc. of Cambridge, Massachusetts.

B. Pharmacology

Various tests on the newer drugs have established their relative efficacy in terms of concentrations required to elicit specific responses, and the results (Razdan *et al.*, 1976) indicate that there is a quantitative difference between the drugs and their effects. For example, in the mouse fighting

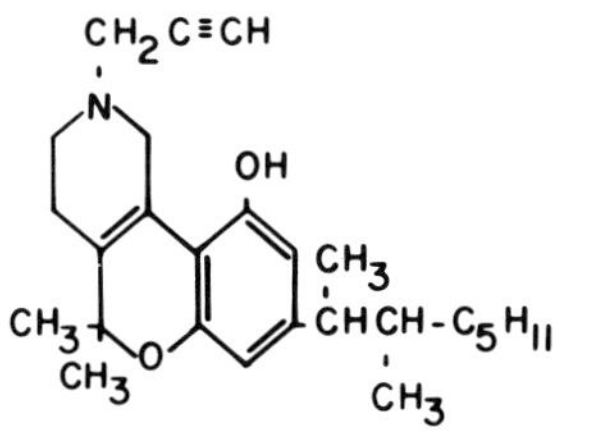

SP—1

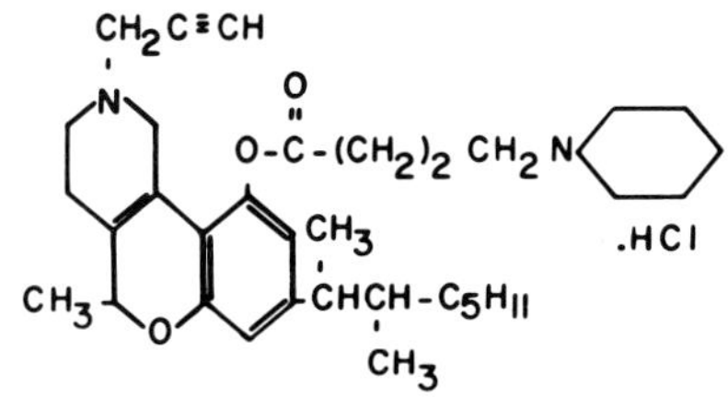

SP—106

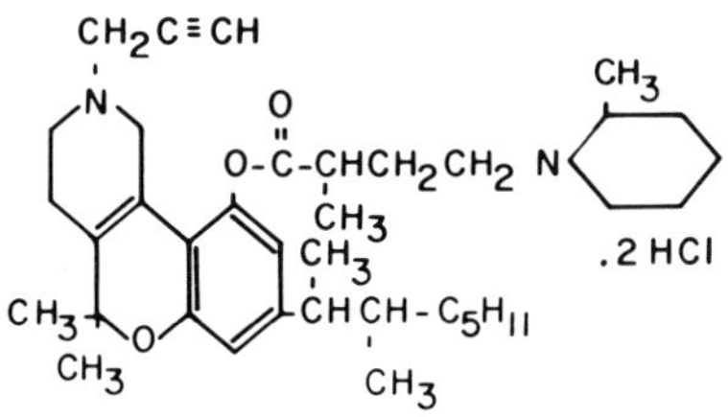

SP—204

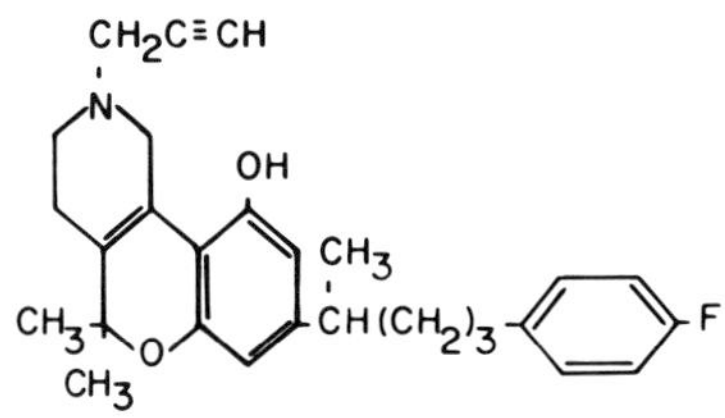

SP—80

FIG. 4. Chemical structure of some new nitrogen-containing cannabinoid analogs.

test, which is a footshock-induced activity, SP-1 was very effective in reducing activity, as was SP-106, but SP-204 was inactive (Razdan *et al.*, 1976). Higher doses of SP-204 and SP-106 than SP-1 were required to induce sedation in adult male cats, and both SP-1 and SP-106 caused a depression in motor activity of rats following treatment whereas SP-204 was less effective. In terms of the analgesic effects of the drugs, however, the order of efficacy was SP-1 > SP-204 > SP-106 (Razdan *et al.*, 1976). It is obvious that structure–activity and dose relationships were not rigidly followed from one test to another. It is this quantitative separation of effects which allows for the development of some new cannabinoid for clinical use in the treatment of elevated IOP, for if a compound can be found which provides a substantial reduction in IOP without causing unwanted side effects, then a major progressive step will have been taken.

C. Ocular Effects

Some initial studies with both SP-1 and SP-106 were made in normal rabbit eyes (Dren *et al.*, 1975; Dren, 1976) which showed that both drugs caused a fall in IOP that lasted for about 4 hr. A concentration of 0.1% caused a fall in IOP of about 23% with the maximum reduction in IOP occurring at 30 min. In another study of the relative toxicity of several new cannabinoids, Thompson and Yang (1976) reported a variety of changes that occurred in dogs with a fluoride-containing derivative, SP-80 (Fig. 4). This drug produced a dose-related sequence of ophthalmic changes that included decreased lacrimation, corneal opacity, keratitis, conjunctivitis, corneal ulceration, and prolapse of the iris. Female dogs treated with 2.8 mg of SP-80/kg/day for 7 days showed 4 hr after the initial treatment a total lack of tear flow that persisted throughout the test period. Recovery of lacrimation was achieved within 24 hr of treatment termination. The IOP was decreased within 1 hr of treatment and lasted for more than 6 hr. The effects on cornea and conjunctiva were most pronounced in female dogs and did not occur in rhesus monkey or rats. The disastrous effects of this drug did tend, therefore, to appear as if they were both species and sex specific (Thompson and Yang, 1976).

In other tests, the effects of SP-1, SP-106, and the fluoride-containing drug (SP-80) were evaluated at oral dose levels of 7.3, 10.8, and 8.0 mg/kg/day in dogs, respectively. All drugs caused a decrease in lacrimation, with both SP-106 and SP-80 causing an almost complete loss of lacrimation; SP-1 was least efficacious in reducing the IOP and both SP-1 and SP-106 caused less conjunctivitis than SP-80. Normal IOP was recovered within 6 hr of treatment, with the maximum fall in IOP being at 1 hr. Body weight loss of 8–9% occurred with all three drugs over the 3-day treatment period, and persisted even after treatment stopped with about a 2-wk recovery period. Behavioral changes, such as drowsiness, ataxia, and inactivity were also noted with all three compounds (Thompson and Yang, 1976).

It should be emphasized, however, that the dose levels employed in this latter study were very high relative to the dose per kilogram body weight required to reduce the IOP in the rabbit (Dren, 1976; Green and Kim, 1976b; Green *et al.*, 1976, 1977a). The dose levels translate, in a 15 kg dog with a plasma volume of 50 ml/kg body weight (Spector, 1956), to plasma concentrations of between 3.8×10^{-3} to 14×10^{-3} mg/ml plasma; values that are almost 1000 times greater than the concentrations required in rabbit and monkey to elicit a substantial fall in IOP (Green and Kim, 1977).

Many of the ophthalmic changes could be the consequence of the absence of lacrimation. The lack of a tear film and subsequent dehydration of the anterior corneal surface will lead to corneal ulceration, conjunctivitis, and keratitis and it would appear that most of the undesirable, ocular drug effects could have been prevented by the use of an adequate, artificial tear solution. With the use of a topically applied drug, however, in a vehicle that itself acts as an artificial tear film such changes may not occur.

Acute dose–response curves were obtained for SP-1, SP-106, and SP-204 in normal albino rabbits and juvenile rhesus monkeys (*Macaca mulatta*) (Green and Kim, 1977). Drugs were applied topically, the lipid-soluble drug (SP-1) in light mineral oil and the two water-soluble drugs (SP-106 and SP-204) in a commercial, artificial tear solution, after a baseline IOP had been determined. The dose–response curves in rabbit are shown in Fig. 5. All drugs caused a fall in IOP compared to the vehicle alone, and the maximal fall occurred at about 2 hr with a subsequent return to baseline IOP at 6 hr with low drug concentrations, and about 8 hr with higher concentrations (Green and Kim, 1977). Drug was applied only in one eye but the contralateral eye also showed approximately the same fall in IOP. Following oral administration through a stomach tube, the fall in IOP was not as marked at equivalent dose levels (on a body-weight basis), which may indicate that there is less absorption through the intestinal wall than occurs through the orbital tissues. The dose–response curve was also less steep which again indicated a lesser absorptive capability when the drug is given orally. The drug effects are of longer duration, however, as well as of slower onset. It is unknown whether the more rapid onset of the fall in IOP caused by a water-soluble drug is caused by better penetration through the mucosal wall of the intestine or better release from the vehicle compared to lipid soluble drugs.

Studies in the rhesus monkey indicated (Green and Kim, 1977) that a

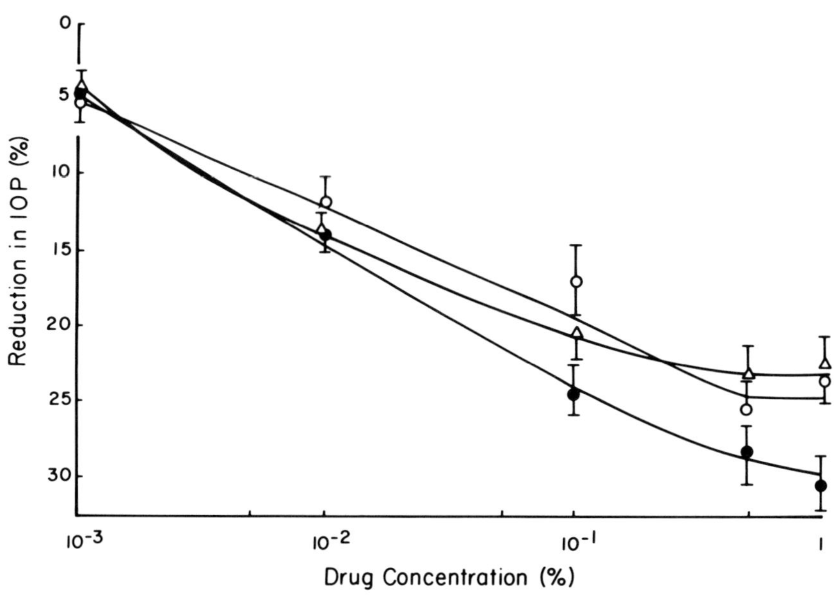

FIG. 5. Reduction in intraocular pressure as a function of concentration of three synthetic cannabinoids. (●) SP-106, (○) SP-1, (Δ) SP-204. All values are the mean ± SEM of at least 6 measurements. (From Green and Kim, 1967a.)

topical concentration of 0.5% is required to cause a fall in IOP similar to that seen in the rabbit with 0.1%. The lower concentration in monkey had no effect on IOP and resembled treatment with vehicle alone. Since species differences have been noted previously with these drugs (Thompson and Yang, 1976) this may well represent another aspect of such interspecies differences. Perhaps the monkey has a greater metabolism of the cannabinoids than the rabbit. Certainly the metabolism of Δ^9-THC differs as regards the time course of conversion of THC to its metabolites between rabbit (Agurell *et al.*, 1970) and man (Wall *et al.*, 1976).

Both SP-1 and SP-106 caused a fall in IOP of normal and ganglionectomized eyes (Green *et al.*, 1977a) although quantitatively different effects were seen between the two drugs. From a comparison of the two drugs it appeared that the fall in IOP caused in the ganglionectomized eye was a greater proportion of the fall in IOP in the normal eye with SP-106 than with SP-1. This implied that SP-106 had more local activity within the eye and, as such, is of importance clinically where more local and less central effects would be desirable. Three of the new cannabinoids, SP-1, SP-106, and SP-204, were all completely blocked with regard to their induction of a fall in IOP by the β-adrenergic antagonist, sotalol, and were partially blocked by the α-adrenergic antagonist, phenoxybenzamine (K. Green and K. Kim, unpublished observations). The new cannabinoids appear, therefore, to be acting in a manner exactly similar to the parent compound, Δ^9-THC (Green and Kim, 1976a).

D. Chronic Treatment

Chronic treatment with SP-1, SP-106, and SP-204 always caused a fall in IOP when the drops were administered topically twice a day (Fig. 6). An initial 9-day study was conducted with the IOP measured every day, and the study was expanded over 60 days with a measurement made every 2 wk (Green *et al.*, 1977a) with further studies continued over the ensuing 10 months (Green *et al.*, 1977c). In all phases of the study, twice-daily application of topical drops of the drugs resulted in a fall in IOP of the normal eye that varied between 15–25% depending on the drug. There was, therefore, no evidence of tolerance since, for each drug, the fall in IOP at the end of 10 months, was the same as it was on the first occasion. No differences were observed in pupil diameter, although there was a trend toward a more miotic pupil (Green *et al.*, 1977b). In addition, a normal weight gain and retention of weight was noted over the 10 month period in the three series of chronic rabbits, each group treated with one of the newer cannabinoids (Green *et al.*, 1977c).

The fall in IOP with systemically (intravenous) and topically adminis-

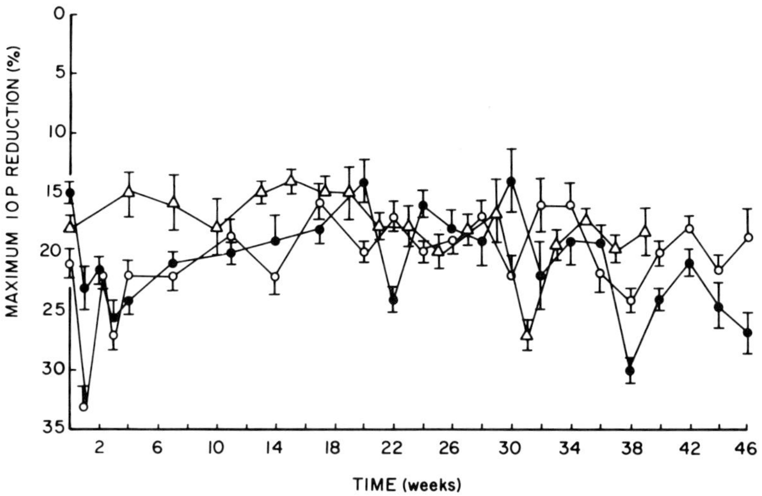

FIG. 6. Reduction in intraocular pressure caused by chronic administration of synthetic cannabinoids. (○) SP-1, (●) SP-106, (Δ) SP-204. All values are the mean ± SEM of at least 9 measurements.

tered SP-106 is approximately the same (Fig. 7), with a slightly better correlation between the two modes of drug delivery than the data obtained with Δ^9-THC (Fig. 3) and may reflect better topical absorption of the water-soluble molecule. The fall in IOP became greater and of longer duration with an absolute maximum fall of about 40% as the concentration of drug was increased. The highest concentration of SP-106 was noticed to cause a reduction in lacrimation when given intravenously, an observation similar to that noted by other workers. No decrease in lacrimation was seen after topical administration of SP-106 even at the highest dose level.

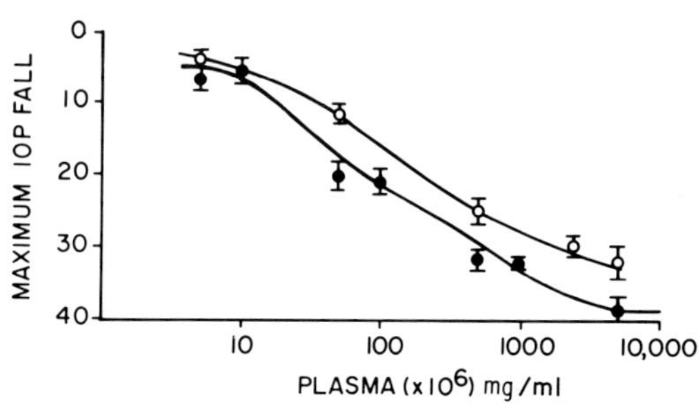

FIG. 7. Comparison of the intraocular pressure fall caused by topical and intravenous administration of SP-106 (●) Intravenous, (○) topical. All values are the mean ± SEM of at least 6 measurements. (From Green *et al.*, 1977b.)

It has been reported that, in man (Hepler *et al.*, 1976) and dog (Thompson and Yang, 1976), lacrimation is decreased after either inhalation of marihuana smoke (Hepler *et al.*, 1976) or oral administration (Thompson and Yang, 1976) of some of the new cannabinoid derivatives. This was not noted where topical drops were used, presumably because of the viscous nature of the vehicles. If decreased lacrimation is significant, the use of a viscous vehicle, such as mineral oil or an artificial tear fluid, for topically applied drugs would tend to counteract this problem.

VIII. FUTURE INVESTIGATIONS

It is obvious from the foregoing that the cannabinoids are effective in causing a reduction in the IOP of several species, including man. The fall in IOP is as good or better than that caused by drugs that are presently available and marketed as antiglaucoma medications (Havener, 1974; Hepler and Petrus, 1976).

Preliminary data in man have indicated that orally administered SP-1 and SP-106 were required at dose levels greater than 0.04–0.15 mg/kg body weight to elicit any noticeable subjective effects (Keats and Romagnoli, 1974) compared to Δ^9-THC where dose levels about one-quarter of those for SP-106 elicited the characteristic subjective high and objective responses of marihuana (Hollister, 1974). The topical and oral doses of SP-1, SP-106, and SP-204 in rabbit and monkey that induce a fall in IOP, are about 0.01–0.08 mg/kg which fall favorably in the range which might be without psychoactive side effects in man. Extrapolation from rabbit to man is unwise, but clinical trials, with dose levels as indicated would quickly answer the question as to whether or not the drugs, when administered in this dose range, lower the IOP.

Obviously, further testing in man is required of some of the most promising compounds to substantiate whether or not the psychoactive properties are reduced. Certainly preclinical testing can identify the structure–activity relationships of many cannabinoids, including other versions of the newer synthetic drugs. The side effects of these drugs need to be assessed in detail, for if the drugs were to be available in the form of a topical drop, it would be hard to convince any agency responsible for allowing drugs to be marketed for public consumption to allow a drug with known psychoactive properties to be released for such public use. The potential would exist for illicit use of the drugs (probably by drinking) but removal, or extremely marked reduction, of the psychoactive properties would render acceptance of the drug an easier hurdle to overcome. Many other factors must also be taken into account, namely the other biological actions of THC and possibly the other cannabinoids; that is, the effects on hormone levels, on the immune system, and tissue repair to name only a few. All of these side effects must be given due and cautious consideration together with the "primary" response of the fall in IOP.

Recent research has indicated that many compounds show at least a quantitative separation of effectiveness in different tests, such that a drug that may be very effective in one test at a low concentration requires a much higher concentration to evoke responses in other measures of activity. What is sought, however, is a drug with full qualitative separation of effects, but this, it must be realized, is an almost impossible criterion to achieve (even for almost all marketed drugs today). Thus, the compound showing the best quantitative effects on IOP with the effects on other measures of activity reduced to a minimum would make such a compound the logical choice for extensive clinical trials. A synthetic cannabinoid would, therefore, seem to be the only logical drug that would achieve clinical acceptance, since molecular manipulation of the compounds might enhance the therapeutic efficacy in the desired area of effect.

It appears that the cannabinoids do potentially offer an alternative to currently used medications in the reduction of elevated IOP. Compounds with little or no side effects or central nervous system activity, with good shelf life and good solubility in appropriate vehicles should all be sought and significant progress has been made toward these end results. A topical preparation offers several advantages over other routes of administration with regard to providing an artificial tear film in the face of decreased lacrimation, allowing a lower dose per application (compared to an oral administration) as well as a lower total quantity of drug per unit package. The low level of toxicity and lack of physical dependence makes the cannabinoids good candidates for clinical use.

The social and legal implications of marihuana ingestion as a drug of abuse have been ignored in this review although certainly this subject is charged with subjective emotions. It would be unfortunate, however, if the potential importance of the cannabinoid and cannabinoid-related drugs were to be influenced, and their development impaired, because of the stigma which is associated with the use of marihuana as a drug of abuse in our society. There is no intent, either real or inferred, on the part of the author to condone the use of marihuana as it is used socially, but as a therapeutically useful drug it is apparent that some compounds based upon the major active ingredient of marihuana show great potential not only as antiglaucoma agents, but also as analgesics, bronchodilators, and antiemetics.

Acknowledgments

This work has been supported in part by Public Health Service Research Grants EY 01413 and EY 01329 from the National Eye Institute and DA 01214 from the National Institute on Drug Abuse. I wish to thank Mrs. K. Bowman, Mrs. K. Kim, and Mr. H. Wynn for their technical assistance and contribution to part of the work reported, and Mrs. I. Prior for her secretarial assistance. Dr. M. C. Braude and Dr. H. Pars generously supplied many of the cannabinoids and the newer synthetic compounds, respectively.

References

Adams, A. J., Flom, M. C., and Jones, R. T. (1972). *Am. J. Optom. Arch. Am. Acad. Optom.* **49,** 880–881.

Adams, R., Pease, D. C., and Clark, T. H. (1940). *J. Am. Chem. Soc.* **62,** 2204–2207.

Adler, C. A., Maurice, D. M., and Paterson, M. E. (1971). *Exp. Eye Res.* **11,** 34–42.

Agurell, S., Nilsson, I. M., Ohlsson, A., and Sandberg, F. (1970). *Biochem. Pharmacol.* **19,** 1333–1339.

Anton, A. H., Serrano, A., Beyer, R. D., and Lavappa, K. S. (1974). *Life Sci.* **14,** 1741–1746.

Banerjee, S. P., Snyder, S. H., and Mechoulam, R. (1975). *J. Pharmacol. Exp. Ther.* **194,** 74–81.

Bárány, E. H. (1963). *Invest. Ophthalmol.* **2,** 584–590.

Bárány, E. H. (1973). *Acta Physiol. Scand.* **88,** 412–429.

Ben-Zvi, Z., and Burstein, S. (1975). *Biochem. Pharmacol.* **24,** 1130–1131.

Ben-Zvi, Z., Mechoulam, R., and Burstein, S. (1970). *J. Am. Chem. Soc.* **92,** 3468–3469.

Blumenthal, M., Yankelev, S., and Dikstein, S. (1976). *Ophthalmol. Res.* **8,** 259–261.

Burstein, S. (1976). *In* "The Therapeutic Potential of Marihuana" (S. Cohen and R. Stillman, eds.), pp. 19–34. Plenum, New York.

Burstein, S., and Raz, A. (1972). *Prostaglandins* **2,** 369–374.

Cavero, I., Buckley, J. P., and Jandhyala, B. S. (1972). *Eur. J. Pharmacol.* **19,** 301–304.

Chiou, C. Y., and Zimmerman, T. J. (1975). *Invest. Ophthalmol.* **14,** 416–417.

Chrai, S. S., Patton, T. F., Mehta, A., and Robinson, J. R. (1973). *J. Pharm. Sci.* **62,** 1112–1121.

Chrai, S. S., Makoid, M. C., Eriksen, S. P., and Robinson, J. R. (1974). *J. Pharm. Sci.* **63,** 333–338.

Coggins, W. J. (1976). *In* "The Pharmacology of Marihuana" (M. C. Braude and S. Szara, eds.), pp. 667–670. Raven, New York.

Cool, S. J., Kaye, S., and Cullen, A. P. (1974). Paper presented at National Association for Research in Vision and Ophthalmology Meeting, Sarasota, Florida, Apr. 25–29, 1974.

Cooler, P., and Gregg, J. M. (1976). *In* "The Therapeutic Potential of Marihuana" (S. Cohen and R. Stillman, eds.), pp. 77–88. Plenum, New York.

Dewey, W. L., Harris, L. S., Howes, J. F., Kennedy, J. S., Granchelli, F. E., Pars, H. G., and Razdan, R. K. (1970). *Nature (London)* **226,** 1265–1267.

Dren, A. T. (1976). *In* "The Therapeutic Potential of Marihuana" (S. Cohen and R. Stillman, eds.), pp. 439–456. Plenum, New York.

Dren, A. T., Bopp, B. A., and Ebert, D. M. (1975). *Pharmacologist* **17,** 267.

Eakins, K. E. (1963). *J. Pharmacol. Exp. Ther.* **140,** 79–84.

Eakins, K. E., and Eakins, H. M. T. (1964). *J. Pharmacol. Exp. Ther.* **144,** 60–65.

Flom, M. C., Adams, A. J., and Jones, R. T. (1975). *Invest. Ophthalmol.* **14,** 52–55.

Frankenheim, J. M., McMillan, D. E., and Harris, L. S. (1971). *J. Pharmacol. Exp. Ther.* **178,** 241–251.

Galanter, M., Wyatt, R. J., Lemberger, L., Weingartner, H., Vaughn, T. B., and Roth, W. T. (1972). *Science* **176,** 934–936.

Gaoni, Y., and Mechoulam, R. (1964). *J. Am. Chem. Soc.* **86,** 1646–1647.

Gill, E. W., and Lawrence, D. K. (1976). *In* "The Pharmacology of Marihuana" (M. C. Braude and S. Szara, eds.), pp. 147–156. Raven, New York.
Gill, E. W., Jones, G., and Lawrence, D. K. (1973). *Biochem. Pharmacol.* **22,** 175–184.
Green, K. (1975). *Invest. Ophthalmol.* **14,** 261–263.
Green, K., and Bowman, K. (1976). *In* "The Pharmacology of Marihuana" (M. C. Braude and S. Szara, eds.), pp. 803–813. Raven, New York.
Green, K., and Downs, S. J. (1975). *Arch. Ophthalmol.* **93,** 1165–1168.
Green, K., and Kim, K. (1975). *Invest. Ophthalmol.* **14,** 36–40.
Green, K., and Kim, K. (1976a). *Invest. Ophthalmol.* **15,** 102–112.
Green, K., and Kim, K. (1976b). *Exp. Eye Res.* **23,** 443–448.
Green, K., and Kim, K. (1977). *Proc. Soc. Exp. Biol. Med.* **154,** 228–231.
Green, K., and Pederson, J. E. (1972). *Am. J. Physiol.* **222,** 1218–1226.
Green, K., and Pederson, J. E. (1973a). *Exp. Eye Res.* **15,** 499–507.
Green, K., and Pederson, J. E. (1973b). *Exp. Eye Res.* **16,** 273–286.
Green, K., and Podos, S. M. (1974). *Invest. Ophthalmol.* **13,** 422–429.
Green, K., Kim, K., and Bowman, K. (1976). *In* "The Therapeutic Potential of Marihuana" (S. Cohen and R. Stillman, eds.), pp. 49–62. Plenum, New York.
Green, K., Bigger, J. F., Kim, K., and Bowman, K. (1977a). *Exp. Eye Res.* **24,** 189–196.
Green, K., Bigger, J. F., Kim, K., and Bowman, K. (1977b). *Exp. Eye Res.* **24,** 197–206.
Green, K., Kim, K., Wynn, H., and Shimp, R. G. (1977c). *Exp. Eye Res.* **25,** 465–471.
Green, K., Wynn, H., and Padgett, D. (1978). *Exp. Eye Res.* **26,** 65–70.
Havener, W. H. (1974). "Ocular Pharmacology." Mosby, St. Louis, Missouri.
Hedqvist, P. (1973). *Adv. Biosci.* **9,** 461–473.
Hedqvist, P., and Brundin, J. (1969). *Life Sci.* **8,** 389–395.
Hepler, R. S., and Frank, I. M. (1971). *J. Am. Med. Assoc.* **217,** 1392.
Hepler, R. S., and Petrus, R. (1976). *In* "The Therapeutic Potential of Marihuana" (S. Cohen and R. Stillman, eds.), pp. 63–76. Plenum, New York.
Hepler, R. S., Frank, I. M., and Ungerleider, J. T. (1972). *Am. J. Ophthalmol.* **74,** 1185–1190.
Hepler, R. S., Frank, I. M., and Petrus, R. (1976). *In* "The Pharmacology of Marihuana" (M. C. Braude and S. Szara, eds.), pp. 815–824. Raven, New York.
Hollister, L. E. (1971a). *Pharmacol. Rev.* **23,** 349–357.
Hollister, L. E. (1971b). *Science* **172,** 21–28.
Hollister, L. E. (1974). *Pharmacology* **11,** 3–11.
Hollister, L. E., and Gillespie, H. K. (1973). *Clin. Pharmacol. Ther.* **14,** 353–357.
Jones, R. T., and Benowitz, N. (1976). *In* "The Pharmacology of Marihuana" (M. C. Braude and S. Szara, eds.), pp. 627–642. Raven, New York.
Keats, A. S., and Romagnoli, A. (1974). Paper presented at *Am. Chem. Soc.*, Atlantic City, New Jersey, Sept. 1974.
Kronfeld, P. C., and Freeman, H. M., (1960). *Am. J. Ophthalmol.* **50,** 1141–1147.
Leopold, I. H. (1955). *Trans. Pac. Coast Oto-Ophthalmol. Soc.* **36,** 43–69.
Leopold, I. H., and Day, R. (1950). *Am. J. Ophthalmol.* **33,** 1452–1453.
Leopold, I. H., and Day, R. (1951). *Arch. Ophthalmol.* **45,** 233–234.
Leopold, I. H., Eisenberg, I. J., and Yasuna, J. (1955). *Arch. Ophthalmol.* **53,** 150–153.
Macri, F. J., and Cevario, S. J. (1974a). *Invest. Ophthalmol.* **13,** 392–395.
Macri, F. J., and Cevario, S. J. (1974b). *Invest. Ophthalmol.* **13,** 617–619.
Mechoulam, R. (1970). *Science* **168,** 1159–1166.
Mechoulam, R., and Edery, H. (1973). *In* "Marihuana" (R. Mechoulam, ed.), pp. 101–133. Academic Press, New York.
Mechoulam, R., and Gaoni, Y. (1965). *J. Am. Chem. Soc.* **87,** 3273–3275.

Mechoulam, R., Shany, A., Edery, H., and Grunfeld, Y. (1970). *Science* **169,** 611–612.

Mechoulam, R., Lander, N., Dikstein, S., Carlini, E., and Blumenthal, M. (1976a). *In* "The Therapeutic Potential of Marihuana" (S. Cohen and R. Stillman, eds.), pp. 35–48. Plenum, New York.

Mechoulam, R., McCallum, N. K., Lander, N., Yagen, B., Ben-Zvi, B., and Levy, S. (1976b). *In* "The Pharmacology of Marihuana" (M. C. Braude and S. Szara, eds.), pp. 39–46. Raven, New York.

Nahas, G. G. (1973). "Marihuana-Deceptive Weed." Raven, New York.

Neufeld, A. H., and Page, E. D. (1975). *Exp. Eye Res.* **20,** 549–561.

Pars, H. G., and Razdan, R. K. (1971). *Ann. N.Y. Acad. Sci.* **191,** 15–22.

Pars, H. G., and Razdan, R. K. (1976). *In* "The Therapeutic Potential of Marihuana" (S. Cohen and R. Stillman, eds.), pp. 419–438. Plenum, New York.

Pars, H. G., Granchelli, F. E., Razdan, R. K., Keller, J. K., Teiger, D. G., Rosenberg, F. J., and Harris, L. S. (1976). *J. Med. Chem.* **19,** 445–454.

Pederson, J. E., and Green, K. (1973a). *Exp. Eye Res.* **15,** 265–276.

Pederson, J. E., and Green, K. (1973b). *Exp. Eye Res.* **15,** 277–298.

Perez-Reyes, M., Timmons, M. C., Lipton, M. A., Davis, K. H., and Wall, M. E. (1972). *Science* **177,** 633–635.

Perez-Reyes, M., Timmons, M. C., Lipton, M. A., Christensen, H. D., Davis, K. H., and Wall, M. E. (1973). *Experientia* **29,** 1009–1010.

Perez-Reyes, M., Wagner, D., Wall, M. E., and Davis, K. H. (1976). *In* "The Pharmacology of Marihuana" (M. C. Braude and S. Szara, eds.), pp. 829–832. Raven, New York.

Poddar, M. K., and Ghosh, J. J. (1976). *In* "The Pharmacology of Marihuana" (M. C. Braude and S. Szara, eds.), pp. 159–173. Raven, New York.

Podos, S. M., Krupin, T., and Becker, B. (1970). *Invest. Ophthalmol.* **9,** 492–495.

Porter, C. D., and Scott, K. G. (1970). *Res. Commun. Chem. Pathol. Pharmacol.* **1,** 733–739.

Purnell, W. D., and Gregg, J. M. (1975). *Ann. Ophthalmol.* **7,** 921–923.

Razdan, R. K., Terris, B. Z., Pars, H. G., Plotnikoff, N. P., Dodge, P. W., Dren, A. T., Kyncl, J., and Somani, P. (1976). *J. Med. Chem.* **19,** 454–461.

Roth, C. D., and Richardson, K. T. (1969). *Am. J. Anat.* **124,** 341–360.

Rubin, V. (1976). *In* "The Therapeutic Potential of Marihuana" (S. Cohen and R. Stillman, eds.), pp. 1–18. Plenum, New York.

Schultes, R. E. (1973). *Nat. Hist.* **82,** 59–68.

Sears, M. L., and Neufeld, A. H. (1975). *Invest. Ophthalmol.* **14,** 83–86.

Shapiro, D. (1974). *Ophthalmologica* **168,** 366–369.

Shaskan, E. G., and Synder, S. H. (1970). *J. Pharmacol. Exp. Ther.* **175,** 404–418.

Snyder, S. H. (1971). "The Uses of Marihuana." Oxford Univ. Press, London and New York.

Sofia, R. D., Ertel, R. J., Dixit, B. N., and Barry, H. (1971). *Eur. J. Pharmacol.* **16,** 257–259.

Spector, W. S., ed. (1956). "Handbook of Biological Data." Saunders, Philadelphia, Pennsylvania.

Sprague, G., and Craigmill, A. (1976). *Res. Commun. Chem. Pathol. Pharmacol.* **14,** 739–742.

Teale, J. D., Clough, J. M., Piall, E. M., King, L. J., and Marks, V. (1975). *Res. Commun. Chem. Pathol. Pharmacol.* **11,** 339–342.

Thompson, G. R., and Yang, C. L. (1976). *In* "The Therapeutic Potential of Marihuana" (S. Cohen and R. Stillman, eds.), pp. 457–474. Plenum, New York.

Vachon, L., and Sulkowski, A. (1976). *In* "The Therapeutic Potential of Marihuana" (S. Cohen and R. Stillman, eds.), pp. 161–172. Plenum, New York.

Vollmer, R. R., Cavero, I., Ertel, R. J., Solomon, T. A., and Buckley, J. P. (1974). *J. Pharm. Pharmacol.* **26,** 186–192.

Walker, R. E., and Litovitz, T. L. (1972). *Exp. Eye Res.* **13,** 14–23.

Wall, M. E., Brine, D. R., Pitt, C. G., and Perez-Reyes, M. (1972). *J. Am. Chem. Soc.* **94**, 8679–8681.

Wall, M. E., Brine, D. R., and Perez-Reyes, M. (1976). *In* "The Pharmacology of Marihuana" (M. C. Braude and S. Szara, eds.), pp. 99–113. Raven, New York.

Whitelocke, R. A. F., and Eakins, K. E. (1973). *Arch. Ophthalmol.* **89**, 494–499.

Willinsky, M. D., Kalant, H., Meresz, O., Endrenyi, L., and Woo, N. (1974). *Eur. J. Pharmacol.* **27**, 106–119.

Zitko, B. A., Howes, J. F., Razdan, R. K., Dalzell, B. C., Dalzell, H. C., Sheehan, J. C., and Pars, H. G. (1972). *Science* **177**, 442–444.

Transport of Organic Solutes in the Lens

H. L. KERN[1]

I.	Introduction	217
II.	Amino Acids	219
III.	Sugars and Polyols	222
IV.	Substituents of Nucleic Acids	225
V.	Effects of Differentiation	226
VI.	Effects of Aging	227
VII.	Effects of Cations, Tonicity, and Chemical Agents	229
VIII.	Effects of Cataractogenic Agents and Cataract	234
IX.	Concluding Remarks	237
	References	238

I. INTRODUCTION

Systems of transport for essential organic substances are present in the plasma membranes of animal cells including the ocular lens. This tissue is of particular interest because of its location in a space free of vasculature, and because of its relatively simple shape, and homogeneous, densely packed population of cells. Additional advantages of the lens for biological studies derive from the fact that it can be excised with a minimum of trauma and will retain its clarity and a reasonably constant metabolism for several hours in a balanced salt–glucose medium. The lens is useful for evaluation of the effects of differentiation because of the separation of the epithelium at the anterior surface from the fiber cells that are formed from this monolayer continually throughout life, and make up the remainder of the tissue.

The systems of transport with which we are concerned are of two types,

[1] Department of Ophthalmology, Albert Einstein College of Medicine of Yeshiva University, Bronx, New York.

ISBN 0-12-153001-9

depending on whether or not there is input of biological energy into the mechanism. Transport of amino acids and *myo*-inositol requires energy and can result in accumulation of compound to a higher level than exists in the medium. Transport of sugars and the substituents of nucleic acids apparently does not require energy in the lens and is of an equilibrating nature (facilitated diffusion).

Entry of solutes into the lens occurs by both saturable and nonsaturable paths—the former presumably requires a proteinaceous carrier situated in the membrane while the latter occurs by unknown means (Christensen and Liang, 1966). Nonsaturable uptake is defined as the residual component of transport, which is linear with concentration when the saturable component(s) is blocked by appropriate competitor(s). Certain small molecules such as water, chloride, urea, and glycerol are believed to traverse the plasma membrane of the lens by so-called simple diffusion which does not involve a carrier. Finally, some substances of relatively low molecular weight, including D-mannitol, most disaccharides, sulfate, and many organic anions are excluded from the cells of the lens.

The transport of amino acids, peptides, and inositol in the lens was reviewed in 1973 by Reddy, who noted that carrier-mediated processes exist for these substances, and that glycolysis can supply the energy required to drive the transport process against the electrochemical gradient. Evidence for the "active" nature of certain of the transport systems listed by Reddy included the accumulation in the lens of nonmetabolizable or slowly metabolizable compounds such as α-aminoisobutyrate (AIB), taurine, and inositol; a requirement for glucose and inhibition by iodoacetate; a Q_{10} of 3–5; inhibition by ouabain, and a rate coefficient for efflux considerably less than that for influx. The importance of the epithelium for the accumulation of amino acids was pointed out. It was also noted that evidence had been obtained for the existence of separate systems for transport of neutral, basic, and acidic amino acids. Specifically, there is experimental support that the A-, L-, Gly-, β- and Ly^+-systems (Christensen, 1975) are present in the lens. The latter system was investigated by Cotlier (1971), who detected in the rabbit lens both saturable and nonsaturable components of uptake of L-lysine, and observed that it was accumulated.

Reference was made by Reddy to the Na^+-dependent nature of the uptake of AIB by the rabbit lens (Cotlier and Beaty, 1967), and to the apparent direct effect of the Na^+,K^+-activated ATPase on the transport of taurine (Reddy, 1970), whereas a slower response to inhibition of this enzyme occurred in transport of the α-amino acids. These observations suggest that transport of amino acids is dependent on the primary active transport of the alkali metal cations. Finally, Reddy discussed evidence obtained in his laboratory for mediated uptake of reduced glutathione (GSH) (Reddy *et al.*, 1973) and glycylglycine (Reddy, 1973), noting that cysteine and alanylglycine were effective inhibitors of the respective systems of transport.[2] No active efflux of oxidized glutathione (GSSG) was detected in the rabbit's lens in contrast to the findings of Srivastava and Beutler (1969) on the human erythrocyte. The latter investigators could find no evidence for exit of GSH from the lenses of rabbits incubated *in vitro*, as only GSSG appeared in the medium (Srivastava and Beutler, 1968).

II. AMINO ACIDS

Our laboratory reported that a mediated transport of dicarboxylic amino acids occurred in the lens of the rat (Kern and Ho, 1974) but not of the calf (Kern and Ho, 1973). The dicarboxylic amino acids appeared to be almost completely excluded from the bovine lens and no competitive effects were observed, indicating the absence of specific sites for transport in plasma membranes of the calf. However, these amino acids evidently do enter the cells of bovine lenses, as Trayhurn and van Heyningen (1973a,b) and Trayhurn (1975) have detected oxidation, incorporation into GSH and protein, and other metabolic alterations of glutamate and aspartate. The transport system in the rat lens had relatively low capacity (v_{max} = 0.11 μmol/gm · hr) and a half saturation coefficient, K_T, of 0.12 m*M* for glutamate. It was rather specific for 4- to 6-carbon L-amino acids, and would accept ω-amides, esters, or sulfonate groups in place of ω-carboxyl groups, though it is not known whether these analogs were transported by this system. The rate coefficient for the nonsaturable component of uptake of glutamate was 0.14 per hr, a value similar to that found for other amino acids in the lens of this species. In contrast to the dicarboxylic amino acids, L-glutamine readily entered the cells of the lenses of rats and calves by a saturable mechanism (Kern and Ho, 1973). When either amino acid was provided in the medium, L-glutamine was incorporated into GSH as the γ-glutamyl residue at least three times as rapidly as L-glutamic acid. Since twice as much glutamine as glutamate was present in the aqueous humor of the calf, it was concluded that glutamine was the physiological precursor of most of the glutamate in the bovine lens. Similar results were obtained by Trayhurn and van Heyningen (1973a) in the lenses of cattle.

We have obtained data supporting the presence in the bovine lens of the ASC-system of Christensen *et al.* (1967; Kern *et al.*, 1977). This system of

[2] Glycylglycine was excluded from the cells of the calf lens (H. L. Kern, unpublished observation).

transport is selective for 3-carbon or 4-carbon neutral amino acids, and is the preferred pathway at physiological levels of L-alanine, L-serine, and L-cysteine. The saturable nature of uptake via the ASC-system of L-serine at both the anterior and posterior faces of the lens (see Section V) is illustrated in Fig. 1. Similar values to those given in this figure were calculated for the kinetic parameters of the ASC-component of L-alanine averaged over the entire lenticular surface (v_{max} = 38 nmol/cm^2 · hr, K_T = 0.14 m*M*). In these experiments, sarcosine was used to block possible uptake by the A-system and the Gly-system (Kern *et al.*, 1977).

Efflux of amino acids from the rat lens evidently occurred by a nonsaturable process (Kern, 1970). This was deduced from the linear nature of exit of AIB and 1-aminocyclopentane-1-carboxylate (ACPC) with concentration of substrate. The rate coefficient of efflux of these amino acids was not significantly different from that for the nonsaturable component of influx. The accelerated exit of these amino acids and of arginine in the presence of an appropriate amino acid in the solution bathing the lens indicated that efflux can occur by a carrier-mediated route. The nonsaturable nature of efflux may be due to competition by the high levels of endogenous amino acids, or it may result from a lower affinity of the amino acids for the transporter functioning in exit as compared to entry. Efflux is difficult to study owing to uncertainties regarding the extent of metabolism and the

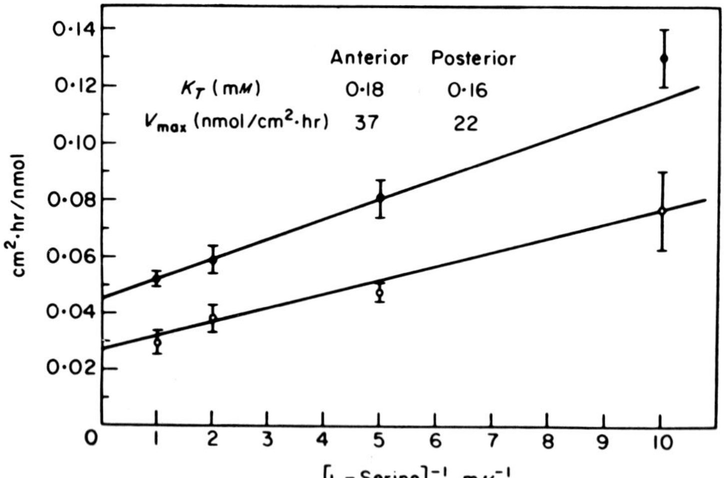

FIG. 1. Kinetics of the ASC-component of L-serine at anterior (○) and posterior (●) surface of calf's lens. Incubation was for 1 hr at 37°, and ASC-component determined as the Na^+-dependent uptake in the presence of fifty times molar excess of sarcosine. (From Kern *et al.*, 1977.)

uniformity of distribution of the radioactive and stable forms of the substrate in the lens.

Guidotti *et al.* (1975) postulated an adaptive control mechanism for transport by the A-system (not L- or Ly$^+$-systems) in embryonic chick lens and in mesenchymal cells from other embryonic and immature tissues. Their data for the lens are summarized in Table I. AIB, L-proline, glycine, L-alanine, and L-serine were shown to have significant uptake by the A-system in embryonic chick heart (Franchi-Gazzola *et al.*, 1973), and they exhibited increased uptake on incubation for 2 hr in Krebs–Ringer-bicarbonate containing 8 m*M* glucose. This effect was not observed for L-phenylalanine or L-lysine, which utilize the L- and Ly$^+$-systems, respectively. Repression–derepression of transport activity by ligands affecting transcription of genes responsible for synthesis of the transporter is thought to be involved. The 20–83% increase in initial rate of uptake resulting from preincubation in the presence of 0.1 m*M* amino acid (Table I) was abolished by preincubation with cycloheximide or 5 m*M* substrate.

A role of intracellular γ-glutamyl peptides, including glutathione, in transport of amino acids or peptides has been proposed (Ball, 1953; Binkley, 1954; Orlowski and Meister, 1970). Evidence has been obtained for the existence in the lenses of rabbits and cattle of the membrane-associated enzyme, γ-glutamyltranspeptidase (Ross *et al.*, 1973; Rathbun and Wicker, 1973), which is postulated to catalyze the reaction of external amino acids or peptides with the γ-glutamyl residue of a cytosolic peptide (initially GSH) to form a γ-glutamyl amino acid or a γ-glutamyl peptide that is subsequently transferred across the plasma membrane.

TABLE I
Amino Acid Uptake by Lens of Embryonic Chick: Changes in Transport with Time[a]

	Uptake (μmol/ml cellular water)		
	Incubation-time in hours		Difference
Amino acid	0	2	(%)
α-Aminoisobutyric acid	0.102	0.187	+83
L-Proline	0.109	0.176	+62
Glycine	0.110	0.139	+26
L-Alanine	0.215	0.338	+57
L-Serine	0.166	0.199	+20
L-Phenylalanine	0.222	0.186	−16
L-Lysine	0.077	0.068	−12

[a] Lenses were obtained from 8-day-old embryos, and four lenses (5–6 mg wet wt.) pooled for analysis. The period of uptake was 10 min. Values are means of 9–15 separate determinations. (From Guidotti *et al.*, 1975.)

The transferred γ-glutamyl peptide can then either be cleaved by γ-glutamylcyclotransferase to yield 5-oxo-L-proline (pyroglutamate) and the free amino acid or peptide, or it can serve as a substrate for the transpeptidase, donating its γ-glutamyl group to another extracellular amino acid or peptide. Therefore, in this scheme, the γ-glutamyl group is required for transport of peptides or amino acids across the cellular membrane. There is no direct evidence that the transpeptidase functions in transport in the lens or in other tissues, though a number of findings indicate it may be involved in translocation of a limited number of neutral amino acids and peptides in jejunum and in kidney (Meister and Tate, 1976; Prusiner *et al.*, 1976). It probably has a minor part to play in transport in the bovine lens since cystine, a good substrate for the enzyme, did not undergo mediated uptake and appeared to be almost completely excluded (H. L. Kern, unpublished observation). In addition, imino acids and α-substituted amino acids are not substrates, but, nonetheless, are readily transported in the lens. There evidently is little transport of α-linked peptides across bovine plasma membranes since, as mentioned, glycylglycine did not enter the calf lens, though this peptide was found by Reddy (1973) to undergo mediated transport in the rabbit lens. However, 10^{-5} *M* bromcresol green, an inhibitor of γ-glutamyltranspeptidase, produced a 50% decrease in the steady state level of taurine, alanine, AIB, methionine, phenylalanine, and glycylglycine in the lens of the rabbit *in vitro* (V. N. Reddy, personal communication). The specificity of this inhibitor needs to be evaluated before the significance of these results is clear.

III. SUGARS AND POLYOLS

Transport of sugars and polyols in the lens has been less extensively investigated than that of amino acids. Kern (1965) presented results indicating saturable uptake of a number of hexoses and of D-xylose, and detected mutual inhibitory effects on entry of several pairs of sugars in the rat lens. D-Mannitol, sodium gluconate, *N*-acetyl-D-hexosamines, and several disaccharides apparently did not penetrate plasma membranes of the calf lens, and D-fructose and D-mannitol were excluded from the cytosol of the mature rat lens. Levari and co-workers (1961) reported that D-fructose was not utilized by the lens of the rat, and Kuck (1962, 1963) observed that the fructose space decreased from 25–40% in young to 10–15% in older rat lenses. Patterson (1965) studied uptake of D-glucose and the nonmetabolizable 3-*O*-methyl-D-glucose (3OMG) in the rat lens, determining that the system was saturable and resembled kinetically that in muscle and human erythrocytes. Elbrink and Bihler (1972) established that there was an equilibrating transport of 3OMG in this species. Their data are shown in Fig. 2, in which the concentration of sugar in the lenticular water is expressed as a percentage of the concentration in the medium. The level of D-glucose was maintained at approximately 1 m*M* throughout the incubation. The methylated sugar had K_T greater than 50 m*M*, its uptake was noncompetitively inhibited by phloretin, and was not significantly Na^+ dependent. Cotlier and Beaty (1967) similarly observed independence of Na^+ for entry of D-galactose, D-glucose, and 3OMG in the rabbit lens. 2-Deoxy-D-glucose was transported as the free sugar in the rodent lens and subsequently phosphorylated, whereas an equilibrating transport of 3OMG appeared to prevail after 8 hr of incubation (Kuck, 1973).

Kern and Ho (1973) presented evidence for comparable mediated trans-

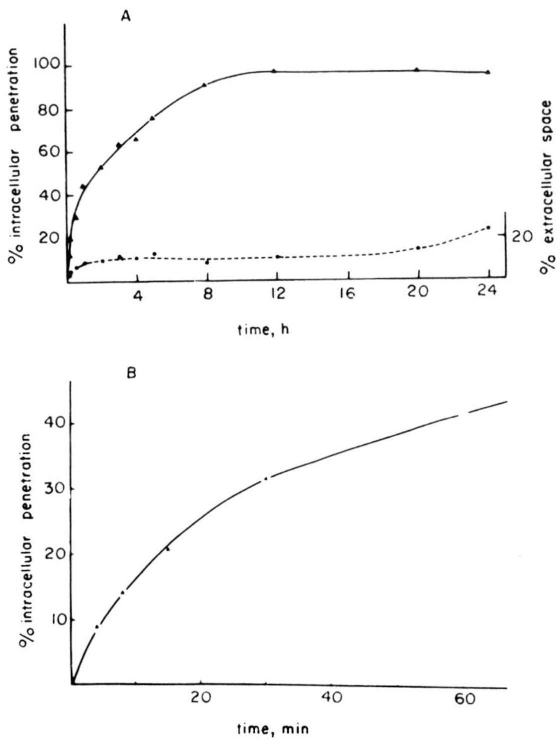

FIG. 2. (A) The time-course of penetration of 5.0 m*M* 3-*O*-methyl-D-glucose (▲) and of inulin (●). Lenses of rats were incubated at 37° in Krebs–Henseleit–bicarbonate buffer containing 1.25 m*M* Ca^{2+}. (B) The time-course of penetration of 3-*O*-methylglucose during the first hour of incubation. (From Elbrink and Bihler, 1972.)

port activity for hexoses at the anterior and posterior faces of the calf lens. A significant nonsaturable component was present at both surfaces, with 2 m*M* phloretin effectively blocking saturable uptake at 0.5–5.0 m*M* levels of sugar. Kinetic analysis showed that the capacity of the system for D-glucose and D-mannose was similar at the two surfaces. Uptake of 3OMG was mediated but linear to 5 m*M* at either surface. Entry of *myo*-inositol, glycerol, and ethylene glycol was not inhibited by D-glucose, though entry of a number of other polyols was. Glycerol and ethylene glycol apparently enter the cells of the lens by simple diffusion. Analysis of transport of a variety of sugars indicated that a C-1 conformation and absence of any group other than H or OH on carbon number one were of primary importance. Recent work in our laboratory suggests that an OH on carbon number one is not essential since 1-anhydro-D-glucitol effectively blocked entry of D-glucose (H. L. Kern, unpublished observation). *meso*-Erythritol was the only polyol having good affinity for the sugar transporter. The order of affinity derived for transport of sugars in the bovine lens closely paralleled that calculated by LeFevre and Marshall (1958) for the human erythrocyte.

A Na^+-dependent, accumulative, mediated transport of *myo*-inositol has been described in the rabbit lens (Cotlier, 1970; Varma *et al.*, 1970) and also exists in the calf lens (H. L. Kern, unpublished observations). Only *scyllo*-inositol and inositol hexaphosphate competed effectively with *myo*-inositol, suggesting that an equatorial orientation of hydroxyl groups about C-1, -2, and -4 was necessary to obtain good affinity for the transporter (Reddy, 1973). It is of interest to compare the structure of D-glucose with that of *myo*- and *scyllo*-inositol (Fig. 3), since these closely related compounds apparently utilize different transporters. It can be seen that all have a chair conformation and an abundance of equatorial hydroxyl groups. Since pentoses in pyranose form utilize the transport system for hexoses (Kern and Ho, 1973), the critical difference in structure appears to be the substitution in the inositol chair of a CHOH group for a ring oxygen, indicating that this oxygen is important in determining the affinity of sugars for the system of transport.

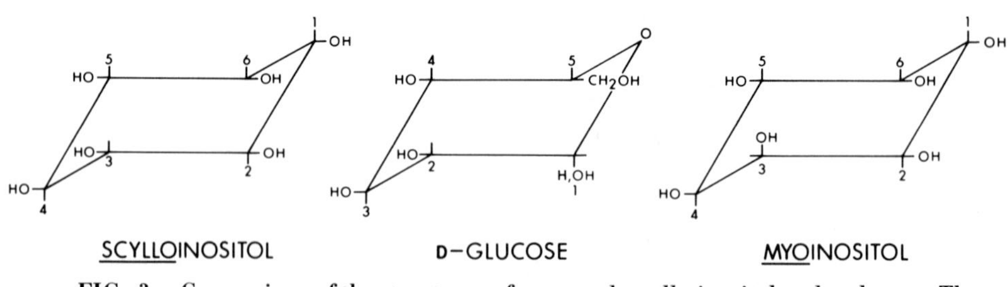

FIG. 3. Comparison of the structures of *myo*- and *scyllo*-inositol and D-glucose. The numbering of the inositols is arbitrary.

IV. SUBSTITUENTS OF NUCLEIC ACIDS

Transport of several constituents of nucleic acids has been examined in the calf lens. The system for nucleosides appeared to be a facilitated diffusion since there was little dependence on external Na^+, and uptake was comparable at the two surfaces as was observed for sugars (Kern and Ho, 1973). Saturable uptake was demonstrated for guanosine, inosine, and adenosine, but no significant uptake of the anionic xanthosine occurred (H. L. Kern, unpublished observations). Nucleosides, and to a lesser degree, purine and pyrimidine bases, inhibited uptake of the ribonucleosides and deoxyadenosine, whereas nucleoside monophosphates and pentoses were ineffective (Tables II and III). The data suggest systems of broad specificity for ribo- and deoxyribonucleosides, but additional work must be done before conclusions can be drawn regarding the number of systems of transport. Uptake of the pyrimidines and of adenine was nonsaturable and unaffected by analogs. However, presumed competitors, including adenine and inosine, were effective against xanthine, hypoxanthine, and guanine, suggesting a separate transport system for oxypurines, although uptake of caffeine was nonsaturable and not inhibited by other oxypurines. It is not known whether the system for oxypurines is distinct from that for nu-

TABLE II
Uptake of 0.1 m*M* Purine Nucleosides by Calf Lens[a]

Nucleoside	Competitor (m*M*)	Uptake (nmol/gm · hr)	Inhibition (%)
	—	15.8 ± 0.5	—
	Adenosine (10)	1.2 ± 0.1	92
	Guanosine (10)	1.6 ± 0.1	90
Inosine	Hypoxanthine (5)	9.8 ± 0.3	41
	5′-IMP (3.8)	13.6 ± 0.2	18
	D-Ribose (5)	14.2 ± 0.3	10
	D-Arabinose (5)	14.5 ± 0.5	8
	—	27.8 ± 0.7, 21.2 ± 0.5	—
	Inosine (10)	8.3 ± 0.1, —	70
Adenosine	Guanosine (10)	13.0 ± 0.3, —	53
	Adenine (5)	26.7 ± 0.5[b]	17
	5′-AMP (5)	— , 21.0 ± 0.7	1
	—	18.8 ± 0.3	—
	Adenosine (5)	4.1 ± 0.1	78
Deoxyadenosine	Deoxyguanosine (5)	6.3 ± 0.1	67
	Deoxyadenosine (5)	5.2 ± 0.1[c]	73
Guanosine	—	22.7 ± 1.0	–
Deoxyguanosine	—	34.0 ± 0.6	–

[a] Incubation was 1 hr at 37°.
[b] Uptake of control was 32.1 ± 1.2 nmol/gm · hr.
[c] Uptake calculated for 0.1 m*M* deoxyadenosine.

TABLE III
Uptake of 0.1 m*M* Pyrimidine Nucleosides by Calf Lens[a]

Nucleoside	Competitor (m*M*)	Uptake (nmol/gm · hr)	Percentage inhibition
Thymine riboside	—	14.0 ± 0.7	—
	Thymidine (5)	3.9 ± 0.2	72
	Uridine (5)	2.7 ± 0.2	80
	Thymine (5)	3.3 ± 0.5	76
Thymidine	—	8.1 ± 0.1	—
	Uridine (5)	2.66 ± 0.07	67
	Deoxyadenosine (5)	3.49 ± 0.10	57
Cytosine arabinoside	—	14.1 ± 1.2	—
	Cytidine (5)	6.8 ± 0.7	52
Deoxycytidine	—	11.3 ± 0.5	—
Uridine	—	19.5 ± 0.8 , 23.6 ± 0.5	—
Deoxyuridine	—	16.9 ± 0.5	—

[a] Incubation was 1 hr at 37°.

cleosides. The entry of 5′-AMP was not saturable, although other nucleoside monophosphates, nucleosides, and adenine were inhibitory. Apparent competitive effects were also noted in uptake of 5′-UMP. Uric acid, orotic acid, and 3′,5′-cyclic AMP were excluded from the cells of the lens.

V. EFFECTS OF DIFFERENTIATION

Reddy remarked in his 1973 review that the role of the lens fiber membranes in transport remained unresolved. The pioneering effort of Kinsey and Reddy in 1965 gave no indication of mediated transport of AIB at the posterior surface of the rabbit lens, nor was there evidence of involvement of carriers in the transport of alkali metal cations or chloride at this surface. In confirmation of this work, Kern and Ho (1973) noted that transport of 0.1 m*M* AIB in the calf lens was about 15 times more rapid at the anterior surface than at the posterior surface, and that methionine did not significantly inhibit uptake at the posterior surface. However, significant mediated uptake of 0.1 m*M* ACPC apparently occurred at the posterior surface, as methionine eliminated 90% of the uptake. The rate of entry of ACPC, a substrate for the L-system, was approximately 10 times as great anteriorly as posteriorly (Kern and Ho, 1973). Data were also presented in this publication supporting the existence at the posterior face of systems of transport for sugars and the nucleoside, thymidine. More recently, Kern *et al.* (1977) did kinetic studies that showed that saturable systems of transport for amino acids are present at the posterior aspect of the calf lens. These include the ASC-, Gly-, and Ly^+-systems (refer to Fig. 1). The coefficients for half saturation, K_T, were comparable at the two surfaces, but v_{max} was less at the posterior surface. On the other hand, no saturable uptake of sarcosine or L-alanine (A-component of uptake) or of ACPC (L-component) was found at the posterior surface, and the rate coefficient of uptake of these amino acids at this face was 5–10% of that at the anterior face. For comparison, the ratio of the rate coefficients, anterior/posterior, of other systems of transport (ASC, Gly, Ly^+, pyranose, and nucleosides) ranged from 0.4–3.4 (Kern *et al.*, 1977). However, our laboratory has recently studied the L-component of L-methionine and detected saturable uptake of this amino acid at the posterior surface. The kinetic data are summarized in Table IV, and show that v_{max} K_T could be determined for L-methionine at the posterior face. A relatively high ratio of the rate coefficients for entry at the two surfaces was obtained as shown in the last column of Table IV. This appears to be characteristic of the A- and L-systems, and is regarded as evidence of their selective alteration on differentiation (Kern *et al.*, 1977). Since a large increase in surface of the cell takes place on differentiation (Kuwabara, 1975), it would appear that certain transporters are synthesized in the fiber cells, but others either are not or are present in a modified, less effective form. These selective changes provide independent support for the existence of discrete systems of transport for amino acids, since the systems to date have only been demonstrated in animal cells by kinetic procedures.

VI. EFFECTS OF AGING

Little information is available concerning the effect of aging on transport in the lens. Sippel, in 1966, showed that the initial rate of uptake of AIB in the lens of the rat declined markedly with age when expressed on a basis of concentration in the tissue or flux across unit area of lenticular surface (Fig. 4). The level of radioactive AIB in lenses of animals 1 yr old was approximately 17% of that in newborn rats. The flux coefficient calculated by Sippel for the 1-month-old rat (0.7 mm/hr) is comparable to that obtained for this amino acid in the calf lens (0.4 mm/hr). On aging, accumulation of 2-deoxy-D-glucose also decreased in the rat lens (Kuck, 1973), as did that of *myo*-inositol in the rabbit lens (Cotlier, 1970). Kern *et al.* (1976) in comparing the transport of amino acids in lenses of calves (2–9 months), steers (1–3 yr) and cows or bulls (6–8 yr) obtained evidence for decreased activity with age. A decline of 55–95% occurred at both surfaces for the ASC-, L-, Gly-, β-, and Ly^+-systems in the cow or bull compared to the calf. The A-system seemed most sensitive to aging since little activity

TABLE IV

Comparison of L-Component of L-Methionine and ACPC at Anterior and Posterior Surfaces of the Calf Lens[a]

Amino acid	V_{max} (nmol/cm^2 · hr)		K_T (mM)		k(mm/hr)		
	Anterior	Posterior	Anterior	Posterior	Anterior	Posterior	A/P
L-Methionine	111	31.2	0.85	1.22	1.07	0.20	5.3
ACPC	83	—	0.17	—	3.10	0.35	8.9

[a] L-Component determined as uptake in 1 hr at 37° of L-methionine or ACPC inhibitable by 100 times molar excess of ACPC or L-methionine, respectively, in a Na^+-free medium in which NaCl was replaced by tetramethylammonium chloride (TEAC) and $NaHCO_3$ by $KHCO_3$. Tonicity was maintained by removal of an amount of TEAC equivalent to the competing amino acid. (From Kern *et al.*, 1977.)

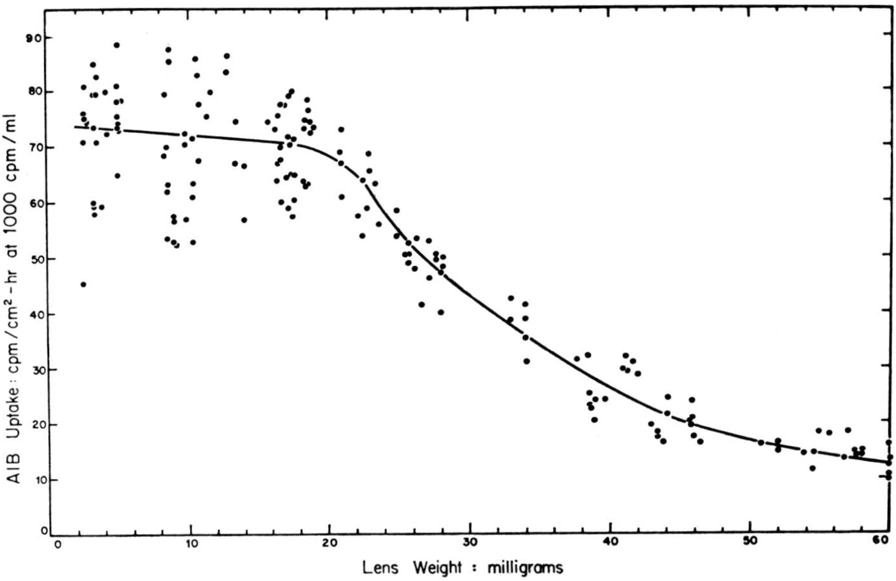

FIG. 4. Uptake of AIB based on surface area in the growing rat's lens. Lenses were incubated 2.5–5.0 hr at 37° in TC199-medium containing 25 m*M* $NaHCO_3$. The values for the ordinate may be converted to flux coefficients in centimeters per hour by dividing by 1000. (From Sippel, 1968.)

was found in the steer and none in the bull. The nonsaturable component of entry of arginine, glycine, and ACPC did not decrease with age. Uptake was expressed per unit of surface area assuming spherical curvature, since this was thought to be the most informative manner of expressing transport in the growing lens. It was concluded that there is either a net loss of transporting molecules (presumably proteins) from the plasma membrane with age, or these molecules are altered so as to become less efficient. The fact that activity of the A-system was lost more rapidly than the others may signify that decreased synthesis of transporters is a factor in the diminution of activity.

VII. EFFECTS OF CATIONS, TONICITY, AND CHEMICAL AGENTS

Calcium is an important factor in many cellular activities (Robertson, 1976; Rasmussen, 1970), and is essential for maintenance of the structural integrity of biological membranes (Gitler, 1976; Montal, 1976). For these reasons, it is worthwhile to assess the functions of this cation in lenticular transport and metabolism. Thoft and Kinoshita (1965) demonstrated that the accumulation of AIB by the rat's lens was inhibited 80% in the absence of Ca^{2+} from the medium, and that efflux of this amino acid was slightly increased. Changes in transport of cations and in extracellular space also occurred. Work in progress indicates that uptake by all systems of transport of amino acids in the calf's lens is depressed 50–85% on omission of Ca^{2+} (Table V). It is apparent from the results in Table V that the β-system and the Gly-system were most drastically altered on removal of this ion. Variable effects were noted on the nonsaturable component—that for taurine, a substrate for the β-system, being increased severalfold, whereas that of ACPC was little affected (Table V).

The deleterious effects of lack of external Ca^{2+} increased with time as illustrated for uptake of 0.1 m*M* L-serine in Fig. 5(A). The rate did not change significantly for the control after preincubation for 2 hr, but that for the test (0 Ca^{2+}) declined from about 73% of the control without preincubation to 48% after 2 hr of preincubation. Figure 5(B) indicates that the extracellular space was markedly increased in Ca^{2+}-free medium in agreement with observations on the rat's lens (Thoft and Kinoshita, 1965). This increase was correlated with an increase in lenticular hydration determined by weighing test and control lenses. Efflux of amino acids utilizing the A-system (AIB, sarcosine) was increased 2–4 times in a medium containing no Ca^{2+}, and there also appeared to be a substantial increase in efflux of taurine, though it was not possible to measure the small cellular leak of this amino acid in the presence of Ca^{2+}. In contrast, efflux of amino acids that use the L-system and the Ly^{+}-system was normal in the deficient medium. Removal of Ca^{2+} decreased the rate coefficient of the β-component for uptake of taurine to about 10% of the control, and saturability could no longer be demonstrated. On the other hand, the kinetic parameters for the Ly^{+}-component of L-arginine were less drastically altered and influx remained saturable (K_T = 160%, v_{max} = 110% of control). The effect on uptake of adenosine and D-glucose was slight on removal of Ca^{2+}, and influx of HTO and thiourea was unchanged (H. L. Kern, unpublished observations). The latter results suggest that simple diffusion is not Ca^{2+} dependent. Elbrink and Bihler (1972) found that transport of D-glucose in the rat lens was uninhibited in a Ca^{2+}-free medium. Similar alterations in transport of cations to those detected by Thoft and Kinoshita were found by our laboratory in the calf lens. Passive fluxes of Na^{+} and K^{+} were increased 5–6 times normal values, but the active fluxes were unchanged. Calcium was required for normal uptake of *myo*-inositol (Cotlier, 1970) and for normal permeability to Rb^{+} in lenses of rabbits (Becker and Cotlier, 1965), and for maintenance of normal levels of alkali metal cations in the bovine lens (Merola *et al.*, 1960). Absence of Ca^{2+} had

TABLE V

Effect of Absence of Ca^{2+} on Systems of Transport of Amino Acids in Calf Lens[a]

0.1 m*M* Amino acid	Transport system[b]	Uptake (nmol/gm · hr) Saturable component $+Ca^{2+}$	Saturable component $-Ca^{2+}$	Saturable component Percentage of control	Nonsaturable component $+Ca^{2+}$	Nonsaturable component $-Ca^{2+}$	Nonsaturable component Percentage of control
ACPC	L	54.3 ± 3.6	26.8 ± 1.8	49	4.65 ± 0.33	3.34 ± 0.24	72
Sarcosine	A	14.5 ± 0.6	4.92 ± 0.28	34	—	—	—
Sarcosine	Gly	22.0 ± 1.0	4.65 ± 0.84	21	3.47 ± 0.27	5.19 ± 0.57	150
L-Arginine	Ly^+	21.2 ± 0.3	8.75 ± 0.21	41	1.70 ± 0.11	7.97 ± 0.48	468
L-Serine	ASC	61.3 ± 4.8	23.3 ± 1.4	38	—	—	—
Taurine	β	2.40[c]	0.38[c]	16	0.41[c]	2.49[c]	607

[a] Incubate 1 hr at 37° following preincubation for ½ hr in Ca^{2+}-containing (control) or in Ca^{2+}-free, TEAC- or Na^+-balanced salt.

[b] Component of uptake determined as described in Table I of Kern *et al.* (1977).

[c] Determined from kinetic plot.

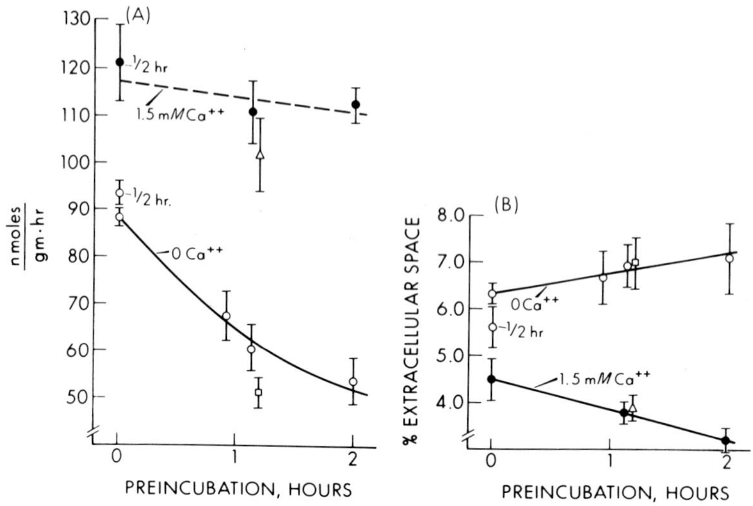

FIG. 5. (A) Effect of time of incubation at 37° in balanced salt–5.0 m*M* D-glucose in the presence or absence of external Ca^{2+} on uptake of 0.1 m*M* L-serine by the calf's lens. The time of incubation in the presence of L-serine was 1 hr (following preincubation for 0–2 hr) except as indicated. The medium was changed only for the points (△,□). (B) Effect of time of incubation with or without Ca^{2+} on the extracellular space expressed as a percentage of the total lenticular water. Conditions of incubation as in (A). The medium was changed only for the points (△,□).

no significant effect on the formation of lactate in 1 hr, but a gradual increase occurred subsequently, and, after 4 hr, it was 57% greater than the control bovine lens (1.73 μmol/gm · hr compared to 1.10 μmol/gm · hr). Extensive structural alterations occurred on incubating this tissue for 2 hr in Ca^{2+}-free balanced salt–glucose. These included hydropic changes in the form of vacuoles and distortion of epithelial cells and superficial processes of the lenticular fibers.

Omission of Mg^{2+} did not appear to significantly disturb transport of the substances mentioned above. However, inhibitory effects of removal of Mg^{2+} on uptake of 2-deoxy-D-glucose and uridine required several hours to develop in embryonic fibroblasts (Bowen-Pope and Rubin, 1977), and proper conditions to demonstrate Mg^{2+} dependence in the bovine lens may not have been achieved. Increasing the external K^+ to 135 m*M* (12 m*M* Na^+) caused a 15% inhibition in uptake of ACPC and L-arginine (H. L. Kern, unpublished observations). These amino acids utilize Na^+-independent systems of transport in the lens. Removal of K^+ inhibited by 0–33% uptake via the L-, A-, Gly-, Ly^+ and ASC-systems. Hypotonicity resulting from elimination of 96 mOsmol/liter of NaCl caused significant inhibition (33–43%) of uptake of ACPC, sarcosine and L-serine by the calf's lens, but did not appreciably alter entry of L-arginine, taurine, HTO, thiourea, Na^+, or Rb^+. However, efflux of Rb^+ was increased threefold in agreement with earlier findings of Cotlier *et al.* (1968). An equivalent hypertonicity had little effect on uptake of amino acids.

We have found that uptake of amino acids is sensitive to sulfhydryl reagents. The relatively nonpenetrating *p*-hydroxymercuriphenylsulfonate (HMPS) is advantageous for studies of transport since its effects can be confined primarily to the cellular surface (Van Steveninck *et al.*, 1965). Table VI illustrates that 1 m*M* HMPS effectively inhibited entry of representative ligands for the six systems of transport for amino acids detected in the calf lens. Uptake of adenosine was inhibited 85%, but there was little effect on entry of D-glucose. We have also observed that 1.0 m*M* iodoacetate and iodoacetamide did not inhibit uptake of D-glucose in 1 hr. Similar observations were made by Elbrink and Bihler (1972) for the hexose transporter in the lenses of rats, and this is in the contrast to the inhibitory effects of sulfhydryl reagents on transport of glucose in human erythrocytes (Van Steveninck *et al.*, 1965; Zipper and Mawe, 1974; Abbott and Schachter, 1976), diaphragm (Battaglia and Randle, 1960), mouse embryo cells (Hare, 1975), and *Neurospora crassa* (Nelson *et al.*, 1975). However, Wilbrandt *et al.* (1947) have reported that iodoacetate did not inhibit uptake of D-glucose in erythrocytes. Transport of amino acids and nucleosides by cultured embryonic cells (Hare, 1975; Hare *et al.*, 1976), *N. crassa* (Nelson *et al.*, 1975), and hepatoma cells (Plagemann and Erbe, 1972) was markedly inhibited by HMPS and other sulfhydryl reagents.

TABLE VI

Effect of 1 m*M* *p*-Hydroxymercuriphenylsulfonate (HMPS) on Uptake of Amino Acids, Adenosine, and D-Glucose[a]

Solute[b]	Transport-system[c]	Uptake nmol/gm · hr	Percent of control
Sarcosine	A	1.51 ± 0.35	10.4
Sarcosine	Gly	0.52 ± 0.22	2.4
L-Arginine	Ly^+	2.04 ± 0.33	9.6
L-Serine	ASC	1.64 ± 0.31	2.7
ACPC	L	0.49 ± 0.09	0.90
Taurine	β	Negligible	<5.0
D-Glucose	Saturable uptake	168 ± 6.4	87
Adenosine	Saturable uptake	4.25 ± 0.14	15

[a] Preincubate for ½ hr in presence of HMPS, change medium and reincubate for 1 hr in presence of HMPS plus solute.

[b] Amino acids and adenosine were 0.1 m*M*; D-glucose, 0.5 m*M*.

[c] Component of uptake determined as described in Table I of Kern *et al.* (1977).

Uptake of ACPC, L-serine, L-arginine, and taurine was inhibited 50–90% in the presence of 0.1 m*M* HMPS, and uptake of adenosine was inhibited 25% (H. L. Kern, unpublished observations). The 85% inhibition of uptake of 0.1 m*M* ACPC by 0.1 m*M* HMPS could be completely reversed by 10 m*M* dithiothreitol, but 10 m*M* reduced glutathione was ineffective. *N*-Ethylmaleimide (10 m*M*) inhibited uptake of 0.5 m*M* D-glucose and 0.1 m*M* adenosine, L-serine and ACPC 63–94%, and 1 m*M* fluorodinitrobenzene, which reacts most readily with amino groups but which may react with –SH groups, inhibited by 69–77% uptake of D-glucose, adenosine and L-serine. Table VII illustrates that the systems of transport are inactivated by 4,4′-dithiodipyridine, a reagent that selectively oxidizes –SH groups (Hare, 1975), but transport of amino acids was little affected by oxidized glutathione (GSSG). This may be because the bulky, polar GSSG cannot gain access to the sulfhydryl group(s) required for activity of the systems transporting amino acids. The specific substrate for glutathione peroxidase, *tert*-butylhydroperoxide, was a relatively ineffective inhibitor of uptake of ACPC, though it was the most effective agent in reacting with epithelial GSH of intact lenses. These data suggest that both sulfhydryl and amino groups may be essential for mediated transport of amino acids, pyranoses, and nucleosides. They also indicate that the reactivity of the sulfhydryl of the transporter for pyranoses differs from that of the other systems of transport, being more resistant to inactivation by poorly penetrating reagents.

A preliminary study was made of two agents that are reactive with receptors and systems of transport of plasma membranes—cytochalasin B (Estensen and Plagemann, 1972; Kletzien *et al.*, 1972) and dipyridamole (Scholtissek, 1968; Kessel and Hall, 1970). Selective effects were noted on the entry of D-glucose, adenosine, and guanosine as shown in Table VIII. Cytochalasin B caused 82% inhibition of the uptake of D-glucose,[3] but had little effect on the nucleosides. In contrast, dipyridamole inhibited by 85–90% the uptake of adenosine, but was ineffective against D-glucose.

VIII. EFFECTS OF CATARACTOGENIC AGENTS AND CATARACT

Several studies have appeared regarding alterations in the transport of amino acids resulting from "sugar" cataracts. Reddy and Kinsey (1963) observed that accumulation of ACPC and AIB was markedly decreased in lenses of alloxan-diabetic rabbits in an investigation that included experi-

[3] In the same experiment, 2 m*M* phloretin inhibited the influx of glucose by 88.5%.

TABLE VII
Effect of 1 m*M* 4,4′-Dithiodipyridine (DTDP), *tert*-Butylhydroperoxide (*T*BHP) and Oxidized Glutathione (GSSG) on Uptake of Amino Acids, Adenosine, and D-Glucose[a]

Solute[b]	Reagent	Transport system	Uptake nmol/gm · hr	Percentage of control
L-Serine	DTDP	ASC	23.6 ± 1.0	41
ACPC	DTDP	Total uptake	11.4 ± 1.3	18
Adenosine	DTDP	Total uptake	5.42 ± 0.62	7.0
D-Glucose	DTDP	Total uptake	71.0 ± 4.8	41
ACPC	TBHP	Total uptake	31.3 ± 2.2	59
L-Serine	GSSG	Total uptake	69.7 ± 2.8	80
ACPC	GSSG	Total uptake	58.7 ± 3.9	96
Sarcosine	GSSG	Total uptake	18.0 ± 0.6	87

[a] Preincubate for ½ hr or 1 hr in presence of reagent or 1% dimethylsulfoxide in controls for lenses incubated in DTDP, and reincubate for 1 hr in presence of reagent plus solute. Preincubate in 10 m*M* GSSG, reincubate in 1 m*M* GSSG.

[b] Amino acids and adenosine were 0.1 m*M*; D-Glucose, 0.5 m*M*.

ments done *in vitro* and *in vivo*. The concentration of many of the naturally occurring amino acids in the lens was also decreased with the exception of arginine, lysine, glycine and, aspartic acid which were at normal or elevated levels. Similarly, lenticular glycine decreased much less than other amino acids in alloxan-diabetic rats according to Patterson *et al.* (1965) and Reddy (1965). Kinoshita *et al.* (1969) found that the levels of the dibasic

TABLE VIII
Effect of Dipyridamole and Cytochalasin B on Uptake by Calf Lens of D-Glucose and Nucleosides

Solute (m*M*)	Inhibitor (μM)[a]	Uptake nmol/gm · hr	Percent inhibition
D-Glucose (1.0)	—	347 ± 26	—
D-Glucose (1.0)	Dipyridamole (10,30)	286 ± 15	17
D-Glucose (0.5)	—	157 ± 7	—
D-Glucose (0.5)	Cytochalasin B (50)	29 ± 1.9	82
Adenosine (0.1)	—	27.5 ± 1.0, 28.5 ± 2.1	—
Adenosine (0.1)	Dipyridamol (10)	– , 4.3 ± 0.35	85
Adenosine (0.1)	Dipyridamole (30)	– , 2.5 ± 0.12	91
Adenosine (0.1)	Cytochalasin B (50)	23.4 ± 0.6, –	15
Guanosine (0.1)	—	19.2 ± 0.6	—
Guanosine (0.1)	Cytochalasin B (50)	16.7 ± 0.9	13

[a] Cytochalasin B was 50 μM for 1 hr preincubation, 10 μM for 1 hr reincubation in presence of solute.

amino acids and glycine were less affected than other amino acids in the lenses of galactose-fed rats. Reddy *et al.* (1976) reported that the concentration of several free amino acids in the lenses of rats, which had been on this diet, increased to 70–80% of normal on returning the animals to a normal diet. This was usually associated with almost complete clearing of the opacity even after the development of mature cataract at 20 days. Increased uptake of glycine and glutamic acid occurred in lenses of rats made diabetic with streptozotocin, although sensitivity to ouabain disappeared according to Laurent *et al.* (1972, 1973) who attributed their findings to increased simple diffusion. Reddy *et al.* (1966) reported that alloxan and pancreatectomy had similar effects on the accumulation of ACPC by lenses of rabbits, indicating that the altered transport was due to the diabetes. These results are suggestive of selective changes in the transport of amino acids in diabetes with the Ly^+- and Gly-systems being less vulnerable than other systems. Cotlier and Beaty (1968) examined the effects of galactose on transport of AIB in lenses of rats and rabbits, and concluded the inhibition was due primarily to the intracellular galactitol formed by the action of aldose reductase or to the increase in hydration caused by the buildup of this polyol. They stated that the level of *myo*-inositol in these lenses was also markedly decreased (see also Kinoshita *et al.*, 1969), and that this substance may play a role in transport of amino acids (Lembach and Charalampous, 1967). Kinoshita *et al.* (1965) pointed out that the deleterious effects of high concentrations of D-glucose or D-galactose *in vitro* on the transport of AIB and ACPC could be largely prevented by adjusting the tonicity of the medium to compensate for the accumulating polyol in the lens. This procedure effectively prevented the net influx of water into the lenticular cells, which evidently leads to a diminished influx and an increased efflux of amino acid.

A lessened rate of entry of AIB was measured by Harris and Gruber (1973) in rats having cataracts induced by triparanol, an inhibitor of synthesis of cholesterol. Transport returned to normal when the compound was removed from the diet. Levels of amino acids were little changed in the cataractous lenses. This cataract is characterized by an early increase in permeability to Na^+ and K^+ (Harris and Gruber, 1972).

The combination of ultraviolet light plus 10 m*M* 3-aminotriazole (an inhibitor of catalase) caused an 85% or greater depression in the steady state levels of AIB or leucine in the lenses of mice, rats, or chicks (Kuck, 1976). Increased hydration, decreased GSH, and decreased incorporation of leucine into protein were also reported. Accumulation of amino acids was not affected by exposure to either agent alone.

Fukui *et al.* (1976) observed a decline in uptake and incorporation of L-leucine as an early effect of hereditary cataract in the Nakano strain of mice. The primary cause of this opacity appears to be an impairment in function of the lenticular Na^+,K^+-activated ATPase, which manifests itself at about the same time as a defect in the transport of leucine (approximately 2 wk after birth). It is not known whether any relation exists between these mechanisms of transport in the Nakano cataract, though an endogenous inhibitor of the ATPase has been identified by these investigators.

IX. CONCLUDING REMARKS

Systems of transport in the lens have been characterized both by kinetic studies that yield data on rate, capacity, and apparent affinity of ligands, and by chemical modification of the solute to obtain information on the specificity of the process. Attempts have been made to examine the systems in isolation by the use of natural or synthetic compounds that have affinity for a known number of classes of transporters (usually one or two), and by taking advantage of the dependence of some of them on external Na^+. This type of analysis indicates that there are six systems of transport for amino acids in the calf lens, whereas only a single system for sugars has been identified. Complexity in transport apparently exists for the constituents of nucleic acids, but the mechanism has not been clearly defined. However, nucleosides enter the lens by a saturable process, and there is evidence that nucleoside monophosphates also undergo mediated transport. A nonsaturable component is present for all organic solutes, which exhibit saturable uptake. It amounts to about 10% of the total cellular influx at physiological levels of ligand. Efflux of amino acids was nonsaturable, having a rate coefficient 50% or less of that for influx, but it could be stimulated by the presence of an appropriate amino acid in the medium. External Ca^{2+} was essential for normal transport of amino acids in the bovine lens, but $Mg^{2+}$was not required in short-term experiments. Removal of Ca^{2+} produced extensive structural damage to lenticular cells after exposure for 2 hr. Transport of amino acids, D-glucose, and nucleosides was inhibited by sulfhydryl reagents, including the relatively nonpenetrating HMPS, and this suggests that proteins in the plasma membrane are the responsible agents. The transporter for D-glucose appears to have a different reactivity from those for the other two classes of compounds, in that it is resistant to attack by anionic sulfhydryl reagents. GSH and GSSG are ineffective *external* reagents possibly because the critical –SH or –S –S – of the transporter is inaccessible to these ionic peptides. There is evidence of selective alteration in transport during differentiation, with drastic decrease in the activity of the A-system, and of marked and possibly selective loss in transport activity on aging of the lens, and in diabetic cataract.

It is evident that considerable insight has been gained concerning the significance and selectivity in transport of the plasma membranes of animal lenses. Nevertheless, there remains a great deal of uncertainty or ignorance regarding these phenomena. More refined kinetic and structural analyses must be done to further define and characterize the macromolecular entities responsible for mediated transport. Little is known concerning compositional differences in the plasma membranes of the epithelial and fiber cells, which might account for differences in transport by these entities, and there is no information regarding the turnover of membrane proteins in the lens. However, several groups of investigators are examining the proteins and lipids of this organelle, and have shown that its isolation and characterization are feasible (Broekhuyse and Kuhlmann, 1974; Benedetti *et al.*, 1976). It is important to attempt to identify and analyze the molecules responsible for transport by fractionation of isolated preparations of lenticular plasma membranes, followed by reconstitution of putative transporters into bilayers of phospholipid, and finally demonstration of function. These are goals that may presently be realized for the relatively abundant fiber cells, if advantage is taken of current biochemical knowledge of biological membranes.

Acknowledgment

The work from this laboratory was supported by grants from Research to Prevent Blindness, Inc., and the National Institutes of Health EY 00268. Portions of this review were presented at the Cataract Workshop of the National Eye Institute in May, 1976.

References

Abbott, R. E., and Schachter, D. (1976). *J. Biol. Chem.* **251,** 7176–7183.

Ahmed, K., and Scholefield, P. G. (1962). *Can. J. Biochem. Physiol.* **40,** 1101–1110.

Ball, E. G., Cooper, O., and Clarke, E. C. (1953). *Biol. Bull. (Woods Hole, Mass.)* **105,** 369–370.

Battaglia, F. C., and Randle, P. J. (1960). *Biochem. J.* **75,** 408–416.

Becker, B., and Cotlier, E. (1965). *Invest. Ophthalmol.* **4,** 117–121.

Benedetti, E. L., Dunia, I., Bentzel, C. J., Vermorken, A. J. M., Kibbelaar, M., and Bloemendal, H. (1976). *Biochim. Biophys. Acta* **45,** 353–384.

Binkley, F. (1954). *In* "Glutathione" (S. Colowick, A. Lazarow, E. Racker, D. R. Schwarz, E. R. Stadtman, and H. Waelsh, eds.), pp. 160–161. Academic Press, New York.

Bowen-Pope, D. F., and Rubin, H. (1977). *Proc. Natl. Acad. Sci. U.S.A.* **74,** 1585–1587.

Broekhuyse, R. M., and Kuhlmann, E. D. (1974). *Exp. Eye Res.* **19,** 297–302.

Christensen, H. N. (1975). *Curr. Top. Membranes Transport* **6,** 227–258.

Christensen, H. N., and Liang, M. (1966). *Biochim. Biophys. Acta* **112,** 524–531.

Christensen, H. N., Liang, M., and Archer, E. G. (1967). *J. Biol. Chem.* **242,** 5237–5246.

Christensen, H. N., de Cespedes, C., Handlogten, M. E., and Ronquist, G. (1973). *Biochim. Biophys. Acta* **300**, 487–522.
Cotlier, E. (1970). *Invest. Ophthalmol.* **9**, 681–691.
Cotlier, E. (1971). *Biochim. Biophys. Acta* **241**, 798–806.
Cotlier, E., and Beaty, C. (1967). *Invest. Ophthalmol.* **6**, 64–75.
Cotlier, E., and Beaty, C. (1968). *Invest. Ophthalmol.* **7**, 77–87.
Cotlier, E., Kwan, B., and Beaty, C. (1968). *Biochim. Biophys. Acta* **150**, 705–722.
Elbrink, J., and Bihler, I. (1972). *Biochim. Biophys. Acta* **282**, 337–351.
Estensen, R. D., and Plagemann, P. G. W. (1972). *Proc. Natl. Acad. Sci. U.S.A.* **69**, 1430–1434.
Franchi-Gazzola, R., Gazzola, G. C., Ronchi, P., Saibene, V., Guidotti, G. G. (1973). *Biochim. Biophys. Acta* **291**, 545–556.
Fukui, H. N., Obazawa, H., and Kinoshita, J. H. (1976). *Invest. Ophthalmol.* **15**, 422–425.
Gitler, G. (1976). *In* "The Enzymes of Biological Membranes" (A. Martonosi, ed.), Vol. V, pp. 229–244. Plenum, New York.
Guidotti, G. G., Gazzola, G. C., Borghetti, A. F., and Franchi-Gazzola, R. (1975). *Biochim. Biophys. Acta* **406**, 264–279.
Hare, J. D. (1975). *Arch. Biochem. Biophys.* **170**, 347–352.
Hare, J. D., Marinetti, G. V., Meisler, A. I., and Tometsko, A. M. (1976). *Biochim. Biophys. Acta* **443**, 485–493.
Harris, J. E., and Gruber, L. (1972). *Invest. Ophthalmol.* **11**, 608–616.
Harris, J. E., and Gruber, L. (1973). *Invest. Ophthalmol.* **12**, 385–388.
Kern, H. L. (1965). *Invest. Ophthalmol.* **4**, 680–690.
Kern, H. L. (1970). *Invest. Ophthalmol.* **9**, 692–702.
Kern, H. L., and Ho, C.-K. (1973). *Exp. Eye Res.* **15**, 751–765.
Kern. H. L., and Ho, C.-K. (1973). *Exp. Eye Res.* **17**, 455–462.
Kern, H. L., and Ho, C.-K. (1974). *Ophthalmic Res.* **6**, 166–174.
Kern. H. L., Ostrove, S. A., and Ho, C.-K. (1976). *Doc. Ophthalmol. Proc.* **8**, 185–192.
Kern. H. L., Ho, C.-K., and Ostrove, S. A. (1977). *Exp. Eye Res.* **24**, 559–570.
Kessel, D., and Hall, T. C. (1970). *Biochim. Biophys. Acta* **211**, 88–94.
Kinoshita, J. H., Merola, L. O., and Hayman, S. (1965). *J. Biol. Chem.* **240**, 310–315.
Kinoshita, J. H., Barber, G. W., Merola, L. O., and Tung, B. (1969). *Invest. Ophthalmol.* **8**, 625–632.
Kinsey, V. E., and Reddy, D. V. N. (1965). *Invest. Ophthalmol.* **4**, 104–116.
Kletzien, R. F., Perdue, J. F., and Springer, A. (1972). *J. Biol. Chem.* **247**, 2964–2966.
Kuck, J. F. R. (1962). *Invest. Ophthalmol.* **1**, 390–394.
Kuck, J. F. R. (1963). *Invest. Ophthalmol.* **2**, 607–611.
Kuck, J. F. R. (1973). *Exp. Eye Res.* **15**, 245–248.
Kuck, J. F. R. (1976). *Invest. Ophthalmol.* **15**, 405–407.
Kuwabara, T. (1975). *Exp. Eye Res.* **20**, 427–443.
Laurent, M., Kern, P., and Regnault, F. (1972, 1973). *Ophthalmic Res.* **4**, 114–121.
LeFevre, P. G., and Marshall, J. K. (1958). *Am. J. Physiol.* **194**, 333–337.
Lembach, K., and Charalampous, F. C. (1967). *J. Biol. Chem.* **242**, 2606–2614.
Levari, R., Kornblueth, W., and Wertheimer, E. (1961). *J. Endocrinol.* **22**, 361–369.
Meister, A., and Tate, S. S. (1976). *Annu. Rev. Biochem.* **45**, 559–604.
Merola, L. O., Kern, H. L., and Kinoshita, J. H. (1960). *AMA Arch. Ophthalmol.* **63**, 830–835.
Montal, M. (1976). *Ann. Rev. Biophys. Bioeng.* **5**, 119–175.
Nelson, S. O., Glover, G. I., and MaGill, C. W. (1975). *Arch. Biochem. Biophys.* **168**, 483–489.

Orlowski, M., and Meister, A. (1970). *Proc. Natl. Acad. Sci. U.S.A.* **67,** 1248–1255.
Patterson, J. W. (1965). *Invest. Ophthalmol.* **4,** 667–679.
Patterson, J. W., Patterson, M. E., Kinsey, V. E., and Reddy, D. V. N. (1965). *Invest. Ophthalmol.* **4,** 98–103.
Plagemann, P. G. W., and Erbe, J. (1972). *J. Cell Biol.* **55,** 161–178.
Prusiner, S., Doak, C. W., and Kirk, G. (1976). *J. Cell. Physiol.* **89,** 853–864.
Rasmussen, H. (1970). *Science* **170,** 404–412.
Rathbun, W. B., and Wicker, K. (1973). *Exp. Eye Res.* **15,** 161–171.
Reddy, D. V. N. (1965). *Invest. Ophthalmol.* **4,** 700–708.
Reddy, V. N. (1970). *Invest. Ophthalmol.* **9,** 206–219.
Reddy, V. N. (1973). *Exp. Eye Res.* **15,** 731–750.
Reddy, D. V. N., and Kinsey, V. E. (1963). *Invest. Ophthalmol.* **2,** 237–242.
Reddy, D. V. N., Kinsey, V. E., and Nathorst-Windahl, G. (1966). *Invest. Ophthalmol.* **5,** 166–169.
Reddy, V. N., Varma, S. D., and Chakrapani, B. (1973). *Exp. Eye Res.* **16,** 105–114.
Reddy, V. N., Schwass, D., Chakrapani, B., and Lim, C. P. (1976). *Exp. Eye Res.* **23,** 483–493.
Robertson, W. G. (1976). *In* "Calcium, Phosphate and Magnesium Metabolism" (B. E. C. Norden, ed.), pp. 230–256. Churchill, London.
Ross, L. L., Barber, L., Tate, S. S., and Meister, A. (1973). *Proc. Natl. Acad. Sci. U.S.A.* **70,** 2211–2214.
Scholtissek, C. (1968). *Biochim. Biophys. Acta* **158,** 435–447.
Sippel, T. O. (1968). *In* "Biochemistry of the Eye" (M. U. Dardenne and J. Nordmann, eds.), Symp. Tutzing Castle, 1966, pp. 277–285. Karger, Basel.
Srivastava, S. K., and Beutler, E. (1968). *Proc. Soc. Exp. Biol. Med.* **127,** 512–514.
Srivastava, S. K., and Beutler, E. (1969). *J. Biol. Chem.* **244,** 9–16.
Thoft, R. A., and Kinoshita, J. H. (1965). *Invest. Ophthalmol.* **4,** 122–128.
Trayhurn, P. (1975). *FEBS Lett.* **50,** 236–238.
Trayhurn, P., and van Heyningen, R. (1973a). *Exp. Eye Res.* **17,** 149–154.
Trayhurn, P., and van Heyningen, R. (1973b). *Biochem. J.* **136,** 67–75.
Van Steveninck, J., Weed, R. I., and Rothstein, A. (1965). *J. Gen. Physiol.* **48,** 617–632.
Varma, S. D., Chakrapani, B., and Reddy, V. N. (1970). *Invest. Ophthalmol.* **9,** 794–800.
Wilbrandt, W., Guensberg, E., and Launer, H. (1947). *Helv. Physiol. Pharmacol. Acta* **5,** C20–C22.
Zipper, H., and Mawe, R. C. (1974). *Biochim. Biophys. Acta* **356,** 207–218.

Subject Index

A

Active transport, in crystalline lens, 73
Aging, effects on lens transport, 227–229
Amino acids, lens transport of, 219–222

B

Benzopyranopyridines, nitrogen-substituted
  chemistry of, 204
  chronic treatment, 208–210
  ocular effects, 205–208
  pharmacology of, 204–205
Binding proteins, for retinol, 143–174
Birds, developing cornea in, collagen origin and distribution in, 1–35

C

Calcium, effects on lens transport, 229–232
Cannabinoids, *See also* individual compounds
  adrenergic interaction of, 196–202
    adrenergic antagonists, 197–199
    ganglionectomy, 199–201
    optic nerve effects, 201–202
    parasympathetic pathways, 201
  chemistry of, 178–181
  clinical ophthalmic tests on, 181–185
  drug delivery and ocular penetration by, 190–193
  intraocular pressure effects of, 181–210
  local and central action of, 202–204
  metabolism of, 179–181
  new types, effects of, 204–210
  nitrogen-substituted, pharmacology, 204–210
  ocular effects of, 175–215
    anterior uveal vessels, 195–196
    glaucomatous eyes, 184
    normal eyes, 181–184
  pharmacology of, 193–202
    theory, 194–195
  preclinical tests on, 185–193
  structure of, 178–179
Cataract and cataractogenic agents, effects on transport in lens, 234–237
Cations, effect on lens transport, 229–234
Chloride, movement of, in lens, 82
Collagen
  aggregation into fibrils, 10–12
  in developing cornea, 1–35
    biochemical studies, 21–25
    cells involved, 12–21
    immunohistochemistry, 25–27
  procollagens for, 5–6
  synthesis and secretion of
    spatial localization, 6–10
  types of, 2–5
Cornea, developing, collagen origin and distribution in, 1–35

E

Extraocular muscle (EOM) fibers
 classification of, 102–107
  global MIFs, 107
  global SIFs, 106
  orbital MIFs, 103
  orbital SIFs, 103
 contractile properties of, 125–140
  tonic system, 132–138
  twitch system, 125–132
 electrical properties of, 108–125
  MIC fibers, 122–125
  multiply innervated fibers, 112–122
  singly innervated fibers, 108–112
 in global region, 93–95
 innervation of, 99–101
 motor nerves of, 100–101
 physiological properties of, 107–140
 sampling problems of, 101–102
 neuromuscular junctions of, 99–100
 in ocular motility, 138–140
 in orbital surface layer, 95–99
  sequential sampling, 96–99
  standard sampling, 95–96
 populations of, 92–99
Eye
 cannabinoid effects on, 175–215
  synthetic compounds, 204–210
 motility of, extraocular fiber role in, 138–140

G

Glaucoma patients, cannabinoid effects on, 184
Golgi apparatus, in collagen secretion, 7–8

H

p-Hydroxymercuriphenylsulfonate, effects on lens transport, 233–234
11-Hydroxy-Δ^9-tetrahydrocannabinol, ocular effects of, 184–185

I

Inositols, lens transport of, 224
Intraocular pressure (IOP)
 cannabinoid effects on, 181–184, 185–211
  pharmacology, 193–202

L

Lens
 amino acid transport by, 219–222
 capsule of, 38–39
 cell-to-cell junctions of, 41–49
  deep fiber, 49
  epithelial, 44–46
  fiber, 46–49
 chloride movement in, 82
 electrophysiology of, 37–90
 as epithelial membrane, 61–66
 epithelium of, 39–40
 extracellular space of, 49–54
  electron microscopy, 49–50
  isotope studies, 50–53
 fiber potential vs. outside layer potential, 54–61
 fibers of, 40–41
 ion distribution and movement in, 71–83
  active transport, 73
  ion distributions, 72–73
  ion flux measurements, 77–83
  potential origin, 83–84
  transport mechanisms, 73–77
 morphology of, 38–49
 nucleic acid component transport by, 225–226
 organic solute transport in, 217–240
  aging effects, 227–229
  cataract and, 234–237
  differentiation effects on, 226–227
  factors affecting, 229–234
 polyol transport by, 222–224
 potassium movement in, 78–80
 resistance measurements of, 66–71
 rubidium movement in, 78–80
 sugar transport by, 222–224
Lipoglycoprotein, retinol-binding, 166

M

Magnesium, effect on lens transport, 232–233
Marihuana, ocular effects of, 175–215
Muscles, extraocular, *see* Extraocular muscles

N

Nerves, of extraocular muscle fibers, 99–101
Nucleic acids, constituents of, lens transport of, 225–226

O

Organic solutes, lens transport of, 217–240

P

Pigment epithelium
  interaction with retinol-binding protein, 153–158
  retinol binding proteins in, 143–174
  role in retinol metabolism, 149–153
Plasma, vitamin A transport in, 145–149
Polyols, lens transport of, 222–224
Potassium, movement of, in lens, 78–80
Procollagens structure and synthesis of, 5–6
Prostaglandin synthetase, cannabinoid inhibition of, 187–188

R

Retina, retinol binding proteins in, 143–174
Retinol
  binding proteins for, *see* Retinol binding proteins
  structure of, 145
  transfer to photoreceptors, 151–153
  uptake from blood, 150–151
    by pigment epithelium, 153–158
Retinol-binding proteins (RBP)
  amino acid composition of, 146, 166
  cytosol proteins, 164–166
  in extraocular tissues, 158–160
  membranc proteins, 160–164
  pigment epithelium reaction with, 153–158
  in photoreceptor cell fractions, 167–169
  plasma transport of, 147–148
Rubidium, movement of, in lens, 78–80

S

Sodium, movement of, in lens, 80–82
SP-derivatives, of cannabinoids, pharmacology of, 204–210
Sugars, lens transport of, 222–224

T

Tear flow, cannabinoid effect on, 181
Δ^8-Tetrahydrocannabinol, ocular effects of, 184–185
Δ^9-Tetrahydrocannabinol
  ocular effects of, 181–184
    glaucomatous, 184
    normal, 181–184
  permeability changes induced by, 186
  preclinical tests on, 185–188
  prostaglandin synthetase inhibition by, 187–188
  structure of, 179

V

"Visual cycle," in frog eye, 152
Vitamin A, plasma transport mechanism for, 145–149